International Workshop on CPT and Lorentz Symmetry in Field Theory 2017

Journal of Physics: Conference Series Volume 952

Faro, Portugal
6 - 7 July 2017

Editors:

Carlos Escobar Ruiz
Jacob Noordmans
Ralf Lehnert
Robertus Potting

ISBN: 978-1-5108-5894-7
ISSN: 1742-6588

Printed from e-media with permission by:

Curran Associates, Inc.
57 Morehouse Lane
Red Hook, NY 12571

Some format issues inherent in the e-media version may also appear in this print version.

Printed by Curran Associates, Inc. (2018)

For permission requests, please contact the Institute of Physics
at the address below.

Institute of Physics
Dirac House, Temple Back
Bristol BS1 6BE UK

Phone: 44 1 17 929 7481
Fax: 44 1 17 920 0979

techtracking@iop.org

Additional copies of this publication are available from:

Curran Associates, Inc.
57 Morehouse Lane
Red Hook, NY 12571 USA
Phone: 845-758-0400
Fax: 845-758-2633
Email: curran@proceedings.com
Web: www.proceedings.com

Table of contents

Volume 952

International Workshop on CPT and Lorentz Symmetry in Field Theory 6–7 July 2017, Faro, Portugal

Accepted papers received: 4 January 2018 Published online: 22 February 2018

Preface

Papers

IOP Conf. Series: Journal of Physics: Conf. Series **952** (2018) 011001 doi:10.1088/1742-6596/952/1/011001

Preface

On 6 and 7 July 2017 the International Workshop on 'CPT and Lorentz Symmetry in Field Theory' took place in Faro, Portugal (`indico.cern.ch/event/596580`). In this workshop we gathered theoretical physicists who are working on or are interested in issues related to CPT and Lorentz symmetry in field theory and gravity. Invariance of the laws of nature under both CPT and Lorentz symmetry is usually taken for granted, and there are profound theoretical reasons behind this idea. Lorentz invariance is generally considered as a basic building block for constructing consistent relativistic quantum field theories. Under the further assumptions of locality and unitarity, the CPT theorem then assures additional invariance of the action under the combined symmetries of charge conjugation, parity inversion, and time reversal. On the other hand, various approaches to quantum gravity allow for departures from CPT and Lorentz symmetry, either spontaneously or otherwise. While at low energies the ensuing effects can be expected to be extremely small due to Planck-scale suppression, they might be detectable in ultra-high precision tests. For these reasons, CPT and Lorentz violation constitute an attractive possibility to look for physics beyond the Standard Model and General Relativity possibly arising from quantum effects in gravity.

Topics of the workshop included theoretical and phenomenological aspects of departures from CPT and Lorentz symmetry, involving field theory, gravitation, and particle physics. The contributions covered a wide range of issues, at the level of fundamental aspects and consistency of field theories such as causality, renormalization and renormalization group, anomalies, stability, effective field theory, (non)commutativity and (non)metricity, as well as more phenomenological issues, including the observed matter-antimatter asymmetry in the Universe, the Casimir effect, and hadronic physics. There were also two presentations about Hořava gravity, including a review talk. The common theme in all of these contributions was always CPT and/or Lorentz symmetry. Taken together, these presentations give a good overview of the state of the art of the field.

The workshop took place in the seminar hall of the School for Hospitality and Tourism of the Algarve, which is located in a historic building in the center of the city of Faro, providing a particularly pleasant and accessible venue for this event. The majority of the workshop participants also took advantage of the Schools hotel located in the same building.

The workshop was opened with welcome addresses by Professor Maria de Lurdes Cristiano, Director of the Faculty of Sciences and Technology of the University of the Algarve, and by Professor Robertus Potting, member of the Organizing Committee. The sharp focus of the workshop made a separation into different topical sessions unnecessary. There were a total of 21 presentations of equal length; the proceedings contributions of these are presented in this volume.

The organizers wish to extend their gratitude to all participants for contributing to an intellectually stimulating atmosphere through their excellent presentations as well as many lively discussions, which made the workshop a success. As editors, we are grateful to all contributors to this proceedings volume for their high-quality papers and for their collaboration in the reviewing process.

The organizers wish to express their gratitude to the School for Hospitality and Tourism of the Algarve for making their premises available for the organization of this workshop. We are grateful to the University of the Algarve and the Multidisciplinary Center for Astrophysics of the Instituto Superior Técnico in Lisbon for organizational and financial support. We thank Ms. Lina Assis Lopes and Ms. Anastasiya Ryabchyk for administrative and secretarial support.

Faro, November 2017

The Editors,

Carlos Escobar Ruíz (University of the Algarve)
Jacob Noordmans (University of the Algarve)
Ralf Lehnert (Indiana University Center for Spacetime Symmetries)
Robertus Potting (University of the Algarve)

Scientific and Organizing Committee

Carlos Escobar Ruíz (University of the Algarve)
Jacob Noordmans (University of the Algarve)
Ralf Lehnert (Indiana University Center for Spacetime Symmetries)
Robertus Potting (University of the Algarve)

Workshop secretariat

Lina Assis Lopes email: `geral@untappedevents.pt`
Anastasiya Ryabchyk

Workshop sponsors

Universidade do Algarve — UAlg (`www.ualg.pt`)
Centro Multidisciplinar de Astrofísica — CENTRA (`centra.tecnico.ulisboa.pt`)
Fundação para a Ciência e a Tecnologia — FCT (`www.fct.pt`)

project number: UID/FIS/00099/2013

IOP Conf. Series: Journal of Physics: Conf. Series **952** (2018) 011001 doi:10.1088/1742-6596/952/1/011001

Participants

Name	Institute
Jean Alexandre	Department of Physics, King's College London, London, UK
Jorge Alfaro	Facultad de Física, Pontificia Universidad Católica de Chile, Santiago, Chile
Orfeu Bertolami	Faculdade de Ciências da Universidade do Porto, Porto, Portugal
Diego Blas	Theoretical Physics Department, CERN, Geneva, Switzerland
Mauro Cambiaso	Departamento de Ciencias Físicas, Universidad Andres Bello, Santiago, Chile
Donald Colladay	New College of Florida, Sarasota, FL, USA
Ian T. Drummond	Department of Applied Mathematics and Theoretical Physics, Centre for Mathematical Sciences, Cambridge, England
Carlos A. Escobar	Departamento de Física, FCT, Universidade do Algarve, Faro, Portugal
Manoel M. Ferreira Jr.	Departamento de Física, Universidade Federal do Maranhão, São Luís, MA, Brazil
Kevin Grosvenor	Niels Bohr Institute, University of Copenhagen, Copenhagen, Denmark
Frans Klinkhamer	Institute for Theoretical Physics, Karlsruhe Institute of Technology, Karlsruhe, Germany
Ralf Lehnert	Indiana University Center for Spacetime Symmetries, Bloomington, IN, USA
Nick E. Mavromatos	Physics Department, King's College London, London, UK
Jacob P. Noordmans	Departamento de Física, FCT, Universidade do Algarve, Faro, Portugal
Jorge Páramos	Faculdade de Ciências da Universidade do Porto, Porto, Portugal
Robertus Potting	Departamento de Física, FCT, Universidade do Algarve, Faro, Portugal
Carlos M. Reyes	Departamento de Ciencias Básicas, Universidad del Bío Bío, Casilla, Chile
Marco Schreck	Departamento de Física, Universidade Federal do Maranhão, São Luís, MA, Brazil
Jay Tasson	Physics and Astronomy Department, Carleton College, Northfield, MN, USA
Luis F. Urrutia	Instituto de Ciencias Nucleares, Universidad Nacional Autónoma de México, Mexico
Zhi Xiao	North China Electric Power University, Beijing, China

IOP Conf. Series: Journal of Physics: Conf. Series **952** (2018) 011001 doi:10.1088/1742-6596/952/1/011001

Group photo

IOP Conf. Series: Journal of Physics: Conf. Series **952** (2018) 011001 doi:10.1088/1742-6596/952/1/011001

Conference poster

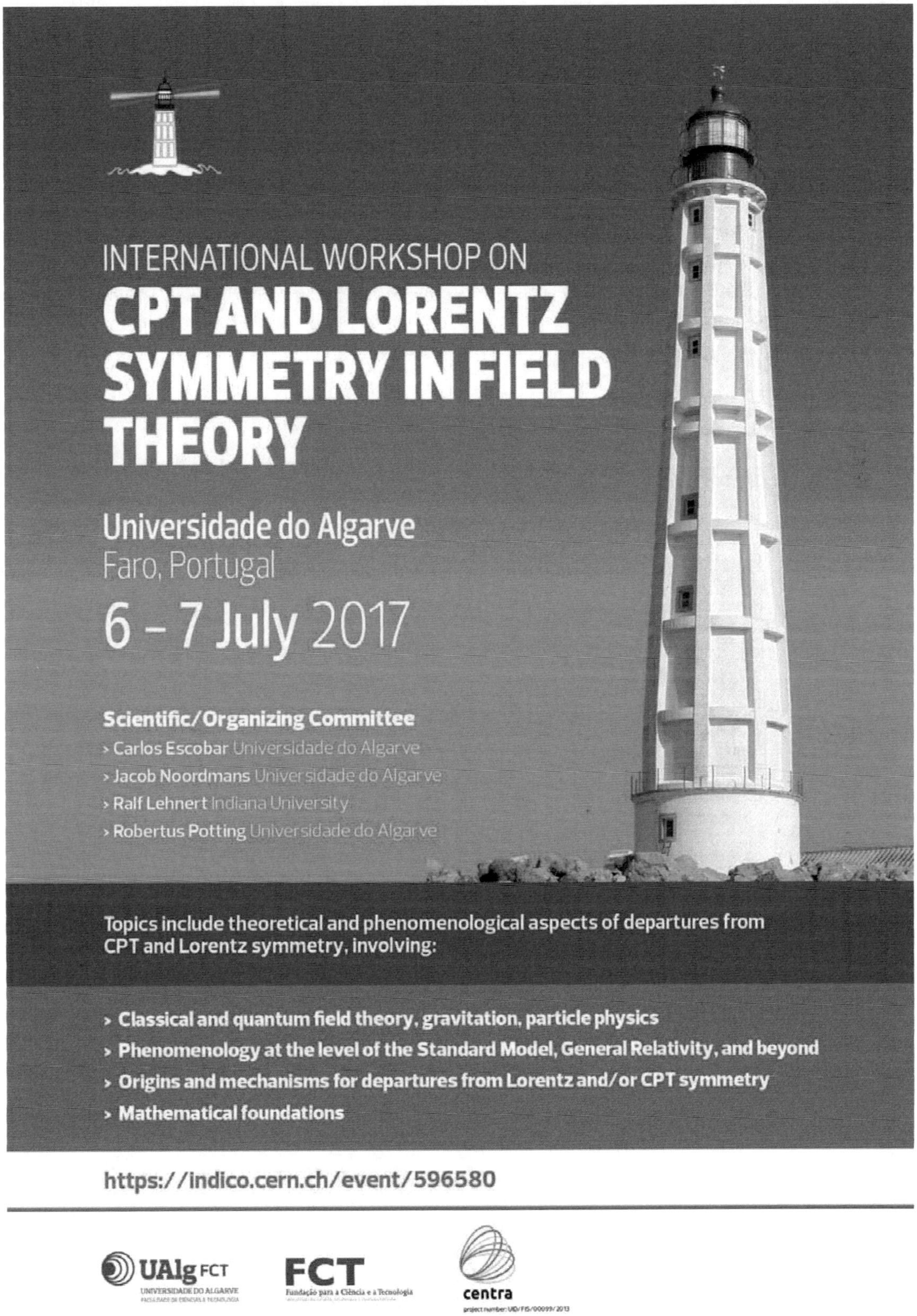

IOP Conf. Series: Journal of Physics: Conf. Series **952** (2018) 011002 doi:10.1088/1742-6596/952/1/011002

Peer review statement

All papers published in this volume of *Journal of Physics: Conference Series* have been peer reviewed through processes administered by the proceedings Editors. Reviews were conducted by expert referees to the professional and scientific standards expected of a proceedings journal published by IOP Publishing.

Lorentz Symmetry Breaking and Causality

I T Drummond

Department of Applied Mathematics and Theoretical Physics, Centre for Mathematical Sciences, Wilberforce Road, Cambridge CB3 0WA, England

E-mail: itd@damtp.cam.ac.uk

Abstract. We investigate Lorentz symmetry (CPT even) breakdown without assuming that it is small. We look for constraints necessary for compatibility with causal structure in the context of scalar theories and QED.

1. Causality with Lorentz invariance

A simple model with Lorentz invariance is a scalar field theory with the action

$$S = \int d^4x \Omega \frac{1}{2}\{g^{\mu\nu}\partial_\mu\phi(x)\partial_\nu\phi(x) + m^2\phi^2(x)\}, \quad (1)$$

with $\Omega = \sqrt{-\det g_{\mu\nu}}$. We assume that the metric is constant throughtout space-time.

An observer is associated with a space-time plane that is space-like with respect to the metric $g_{\mu\nu}$. Time evolution is associated with a space-like foliation of space-time. The foliations are given a time label by means of a time-like vector n^μ. This is illustrated in figure 1. Under Lorentz transformations $g_{\mu\nu}$ is invariant as is the associated light-cone. The space-like foliation is changed corresponding to the associated change of observer.

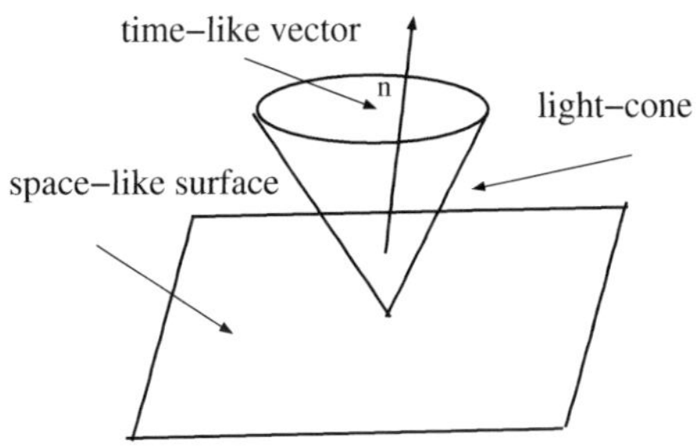

Figure 1. Example of a light-cone with space-like surface and time-like evolution direction.

2. Lorentz symmetry breaking

A model that exhibits Lorentz symmetry breaking is one with two fields [1] governed by an action

$$
\begin{aligned}
S &= \int d^4x\Omega\frac{1}{2}\{g^{\mu\nu}\partial_\mu\phi(x)\partial_\nu\phi(x) + m^2\phi^2(x)\} \\
&\quad + \int d^4x\Omega\frac{1}{2}\{\bar{g}^{\mu\nu}\partial_\mu\psi(x)\partial_\nu\psi(x) + \bar{m}^2\psi^2(x)\},
\end{aligned} \tag{2}
$$

We are free to set $\det \bar{g}_{\mu\nu} = -\Omega^2$. If the two metrics are not identical then the model exhibits a breaking of Lorentz symmetry but with no restriction on its magnitude. Figure 2 indicates possible relationships between the two light-cones that are consistent with causal propagation because the two metrics share foliations of space-time that are space-like in both.

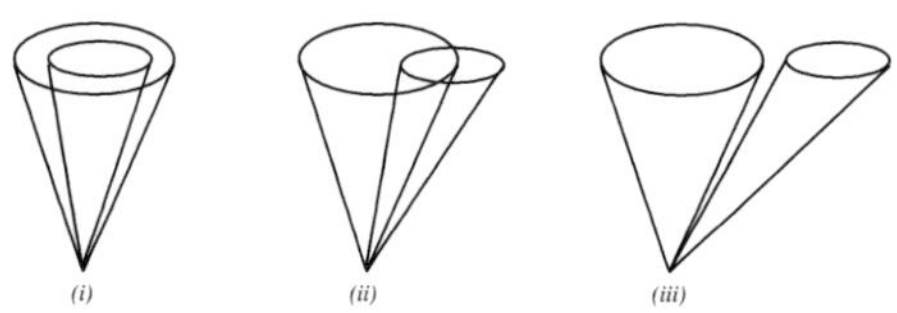

Figure 2. Examples of pairs of light-cones that share space-like surfaces. The first two examples represent situations in which there are space-time vectors that are time-like in both cones. In the third case each light-cone is seen as space-like from the other.

It is formally possible for the two light-cones to be so rotated with respect to each other that there are no foliations of space-time that are space-like with respect to both metrics. We exclude such circumstances on the grounds that they do not permit simultaneous causal propagation of the two fields.

There have been many careful studies of the breakdown of Lorentz invariance [2, 3, 4] with the conclusion that observations constrain it to be a very small effect indeed [5]. Nevertheless it is interesting to consider what issues of principle are involved in dealing with quantum field theory in the context of a multi-metric space-time without any *a priori* constraint on the relationship of the various metrics except for the general requirement of maintaining a causal structure for the theory. These issues of causality and multiple lightcone structure appear in other contexts [6, 7].

3. Mixing interaction

We encounter further restrictions on the relationship between the two metrics when we include an interchange term in the action [1] of the form

$$
S_I = \int d^4x\Omega\{w^2\phi(x)\psi(x)\}. \tag{3}
$$

The equations of motion become

$$
\begin{aligned}
g^{\mu\nu}\partial_\mu\partial_\nu\phi(x) + m^2\phi(x) + w^2\psi(x) &= 0, \\
\bar{g}^{\mu\nu}\partial_\mu\partial_\nu\psi(x) + \bar{m}^2\psi(x) + w^2\phi(x) &= 0.
\end{aligned} \tag{4}
$$

Plane wave solutions are $\phi(x) = Ae^{-iq_\mu x^\mu}$ and $\psi(x) = \bar{A}e^{-iq_\mu x^\mu}$, where

$$
\begin{aligned}
(g^{\mu\nu}q_\mu q_\nu - m^2)A - w^2\bar{A} &= 0, \\
(\bar{g}^{\mu\nu}q_\mu q_\nu - \bar{m}^2)\bar{A} - w^2 A &= 0.
\end{aligned} \tag{5}
$$

For consistency we have the determinantal condition which yields the dispersion relation

$$
(g^{\mu\nu}q_\mu q_\nu - m^2)(\bar{g}^{\mu\nu}q_\mu q_\nu - \bar{m}^2) - w^4 = 0. \tag{6}
$$

In the marginal case $w^2 = m\bar{m}$ this becomes for low q_μ

$$(\bar{m}^2 g^{\mu\nu} + m^2 g^{\mu\nu}) q_\mu q_\nu = 0. \tag{7}$$

The effective light-cone at low q_μ has the form

$$\hat{g}^{\mu\nu}(u) q_\mu q_\nu = 0, \tag{8}$$

where $\hat{g}^{\mu\nu}(u) = u g^{\mu\nu} + (1-u)\bar{g}^{\mu\nu}$ with $u = \bar{m}^2/(m^2+\bar{m}^2)$. Hence $\hat{g}^{\mu\nu}(u)$ interpolates between the two original metrics. To maintain causality it is important that $\det \hat{g}^{\mu\nu}(u)$ does not vanish and change sign for $0 \leq u \leq 1$. This requires the two original metrics not only to share space-like foliations but in addition to have in common some time-like vectors. Geometrically this means that example *(iii)* in figure 2 is ruled out. Detailed arguments are presented in Ref. [1].

4. Quartic interaction

The above argument for the interchange interaction is suggestive rather than rigorous. However the same issue must be addressed in a model involving quartic interactions that include the term

$$S_I = \int d^4x\Omega \left\{ -\frac{1}{4}\sigma\phi^2(x)\psi^2(x) \right\}. \tag{9}$$

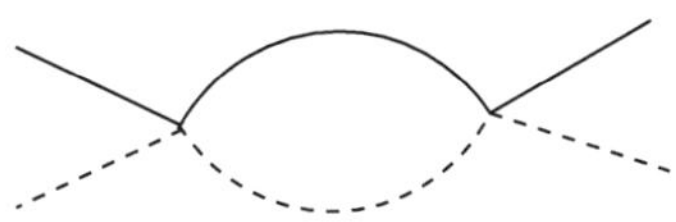

Figure 3. Loop diagram resulting from the $\phi^2\psi^2$ interaction. Solid lines correspond to the ϕ-propagator and dashed lines to the ψ-propagator.

In the computation of Green's functions this leads to a Feynman diagram of the form shown in figure 3. Using the Schwinger representation of field propagators we obtain (at zero momentum) the result

$$\int d^n k \int_0^1 du \int_0^\infty d\lambda\lambda \exp\{i\lambda(\hat{g}^{\mu\nu}(u) - um^2 - (1-u)\bar{m}^2)\},$$

where $\hat{g}^{\mu\nu}(u)$ is given above. This result is proportional to

$$\int_0^1 \frac{du}{\sqrt{-\det \hat{g}_{\mu\nu}(u)}} (um^2 + (1-u)\bar{m}^2)^{n/2-2}.$$

We see immediately in this case we must require that $\det \hat{g}^{\mu\nu}(u) \neq 0$ in the range $0 \leq u \leq 1$ in order that the result is well defined.

5. Lorentz symmetry breaking in QED - photons

The most general gauge invariant action for electrodynamics is the pre-metric form [8]

$$S = \int d^4x U^{\mu\nu\sigma\tau} F_{\mu\nu}(x) F_{\sigma\tau}(x). \tag{10}$$

Here

$$F_{\mu\nu}(x) = \partial_\mu A_\nu(x) - \partial_\nu A_\mu(x). \tag{11}$$

IOP Conf. Series: Journal of Physics: Conf. Series **952** (2018) 012001 doi:10.1088/1742-6596/952/1/012001

The tensor density $U^{\mu\nu\sigma\tau}$ has the symmetry properties of the Riemann tensor.

$$U^{\mu\nu\sigma\tau} = -U^{\nu\mu\sigma\tau} = U^{\sigma\tau\mu\nu}. \tag{12}$$

$$U^{\mu\nu\sigma\tau} + U^{\mu\sigma\tau\nu} + U^{\mu\tau\nu\sigma} = 0. \tag{13}$$

Standard Lorentz invariant electrodynamics is associated with an (invariant) metric $g^{\mu\nu}$ for which

$$U^{\mu\nu\sigma\tau} \propto g^{\mu\sigma}g^{\nu\tau} - g^{\nu\sigma}g^{\mu\tau}. \tag{14}$$

Although pre-metric theory has apparently no associated metric, it does in fact have a *preferred metric* - $g_{\mu\nu}$ [9]. It can be defined by the requirement that it makes $g^{\mu\sigma}g^{\nu\tau} - g^{\nu\sigma}g^{\mu\tau}$ "closest to" $U^{\mu\nu\sigma\tau}$. We realise this idea by demanding that the overlap $F = U^{\mu\nu\sigma\tau}(g_{\mu\sigma}g_{\nu\tau} - g_{\nu\sigma}g_{\mu\tau})$ is stationary with respect to variations of $g_{\mu\nu}$ subject to the constraint $\det g_{\mu\nu} = -\Omega^2$. Introducing the Lagrange multiplier λ, we have

$$\delta(F - \lambda \det g_{\mu\nu}) = 0, \tag{15}$$

with as a result the eigenvalue-like equation [10]

$$U^{\mu\nu\sigma\tau}g_{\nu\tau} + \lambda\Omega^2 g^{\mu\sigma} = 0. \tag{16}$$

There is a standard decomposition of a tensor with the symmetry properties of $U^{\mu\nu\sigma\tau}$ namely

$$\begin{aligned} U^{\mu\nu\sigma\tau} &= \frac{U}{12}(g^{\mu\sigma}g^{\nu\tau} - g^{\nu\sigma}g^{\mu\tau}) \\ &+\frac{1}{2}(U^{\mu\sigma}g^{\nu\tau} + U^{\nu\tau}g^{\mu\sigma} - U^{\nu\sigma}g^{\mu\tau} - U^{\mu\tau}g^{\nu\sigma}) \\ &-C^{\mu\nu\sigma\tau}, \end{aligned} \tag{17}$$

where $U = U^{\mu\nu\sigma\tau}g_{\mu\sigma}g_{\nu\tau}$ and $U^{\mu\sigma} = U^{\mu\nu\sigma\tau}g_{\nu\tau} - (1/4)Ug^{\mu\sigma}$. The residual tensor $C^{\mu\nu\sigma\tau}$ has the symmetry properties of $U^{\mu\nu\sigma\tau}$. In addition it is traceless,

$$C^{\mu\nu\sigma\tau}g_{\nu\tau} = 0. \tag{18}$$

That is $C^{\mu\nu\sigma\tau}$ has the properties of the Weyl tensor. We refer to it as a Weyl-like tensor (WLT). In the present case for our preferred metric we see that $U^{\mu\sigma} = 0$. so this decomposition reduces to

$$U^{\mu\nu\sigma\tau} = \frac{U}{12}(g^{\mu\sigma}g^{\nu\tau} - g^{\nu\sigma}g^{\mu\tau}) - C^{\mu\nu\sigma\tau}, \tag{19}$$

We can therefore use the standard Petrov classification of Weyl tensors [11] to classify Lorentz symmetry breaking in electrodynamics. We note that the gauge invariant equations of motion, for an appropriate normalisation of $U^{\mu\nu\sigma\tau}$, take the form

$$(g^{\mu\sigma}g^{\nu\tau} - g^{\mu\tau}g^{\nu\sigma} - C^{\mu\nu\sigma\tau})\partial_\mu\partial_\sigma A_\nu(x) = 0. \tag{20}$$

6. Petrov classification of Weyl tensors

The Petrov classification, see Refs. [12, 13], is based on light-like vectors v_μ that satisfy

$$v_{[\alpha}C_{\mu]\nu\sigma[\tau}v_{\beta]}v^\nu v^\sigma = 0. \tag{21}$$

The 4-vector v^μ is referred to as a *principal null direction* - PND. There are four PNDS that are solutions though they may coincide, with various multiplicities. The geometrical significance of

IOP Conf. Series: Journal of Physics: Conf. Series **952** (2018) 012001 doi:10.1088/1742-6596/952/1/012001

the PNDs is most easily grasped using the Newman-Penrose spinor formalism but this requires a whole separate presentation see Refs. [12, 13]. More elementary accounts of this approach can be found in Refs. [14, 15].

The Petrov classification is based on the number and multiplicities of the PNDs. The extra conditions satisfied by PNDs of higher multiplicity can be found in Refs. [12, 13].

Four distinct PNDs	-	Class I
Two distinct PNDs each with multiplicity 2	-	Class D
Three distinct PNDs one with multiplicity 2	-	Class II
Two distinct PNDs with multiplicities 3 and 1	-	Class III
A single PND with multiplicity 4	-	Class N

6.1. Newman-Penrose tetrad

The Newman-Penrose tetrad is useful for analysing the structure of WLTs. It comprises four null vectors l_μ, n_μ (real), m_μ, $\bar{m}_\mu$ (complex conjugate). The only nonvanishing scalar products are

$$l^\mu n_\mu = 1 = -m^\mu \bar{m}_\mu.$$

Hence

$$g_{\mu\nu} = l_\mu n_\nu + n_\mu l_\nu - m_\mu \bar{m}_\nu - \bar{m}_\mu m_\nu.$$

6.2. Canonical forms - simplest case

Each Petrov class can be given a canonical form that reveals the structure of the WLT $C^{\mu\nu\sigma\tau}$. The simplest case is class N. Choose the PND to be l_μ of the NP tetrad and construct the anti-symmetric tensor $A_{\mu\nu} = l_\mu m_\nu - m_\mu l_\nu$. The canonical form is

$$C_{\mu\nu\sigma\tau} = \lambda A_{\mu\nu} A_{\sigma\tau} + \text{c.c.} \tag{22}$$

We can choose λ to be real. Other classes have more complicated canonical forms.

For plane wave solutions $A_\mu(x) = \epsilon_\mu e^{-iq_\sigma x^\sigma}$ the gauge invariant equations of motion are, after imposing the gauge condition $q^\mu \epsilon_\mu = 0$,

$$(q^2 g^{\tau\nu} - C^{\mu\nu\sigma\tau} q_\mu q_\sigma)\epsilon_\nu = 0. \tag{23}$$

For Petrov class N this becomes

$$(q^2 g^{\tau\nu} - \lambda(Q^\tau Q^\nu + \bar{Q}^\tau \bar{Q}^\nu))\epsilon_\nu = 0. \tag{24}$$

where $Q^\nu = A^{\mu\nu} q_\mu$. Using the results $Q^2 = \bar{Q}^2 = 0$, and $Q.\bar{Q} = -(l.q)^2$ we find

$$\begin{pmatrix} q^2 & \lambda(l.q)^2 \\ \lambda(l.q)^2 & q^2 \end{pmatrix} \begin{pmatrix} Q.\epsilon \\ \bar{Q}.\epsilon \end{pmatrix} = 0. \tag{25}$$

The consistency of these equations requires the determinantal condition $(q^2)^2 - (\lambda(l.q)^2)^2 = 0$ which factorises to yield two light-cones $g^{(\pm)\mu\nu} q_\mu q_\nu = 0$, where

$$g^{(\pm)\mu\nu} = g^{\mu\nu} \pm \lambda l^\mu l^\nu.$$

The two light-cones correspond to different photon polarisations - hence birefringence. It is easy to check that the interpolating metric $\hat{g}^{\mu\nu}(u) = u g^{(+)\mu\nu} + (1-u) g^{(-)\mu\nu}$ is never singular. The

IOP Conf. Series: Journal of Physics: Conf. Series **952** (2018) 012001 doi:10.1088/1742-6596/952/1/012001

light-cone structure is therefore consistent with causality in straightforward manner. See also Ref. [16].

The quartic dispersion relation for other Petrov classes may not factorise into two quadratic light-cones see Ref. [9]. For example class III (see Ref. [9]) yields

$$(q^2 - (l.q)^2)^2 - (l.q)^2(4m.q - l.q)(4\bar{m}.q - l.q) = 0.$$

Here the photon propagator does not have a Schwinger-style representation which results in further complication in analysing the causal structure in perturbation theory in QED.

7. QED - photons and electrons

QED requires us to introduce the electron-photon interaction and in particular to evaluate the vacuum polarisation diagram in figure 4.

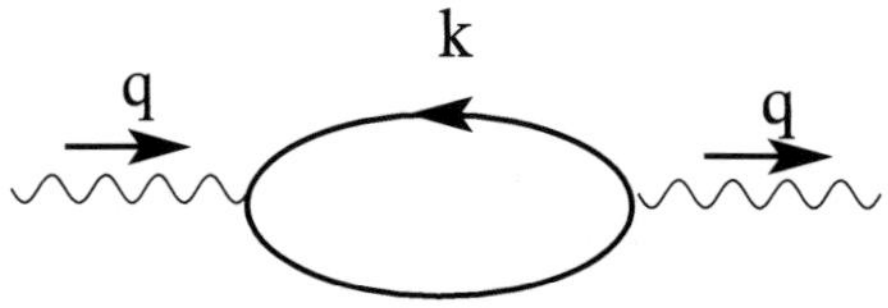

Figure 4. Vacuum polarisation diagram in QED.

If Lorentz symmetry breaking arises because of a different metric, $\bar{g}^{\mu\nu}$, for the electron then the vacuum polarisation tensor has a structure that reflects this new metric and exhibits a UV divergence of the form

$$\Sigma^{\mu\nu}(q) = \frac{e^2}{6\pi^2}\frac{1}{n-4}W^{\mu\alpha\nu\beta}q_\alpha q_\beta,$$

where

$$W^{\mu\alpha\nu\beta} = \bar{g}^{\mu\nu}\bar{g}^{\alpha\beta} - \bar{g}^{\mu\beta}\bar{g}^{\nu\alpha}.$$

The tensor coefficient of the UV divergence can be decomposed in the standard way to reveal a contribution $V^{\mu\alpha\nu\beta}$ that is a WLT,

$$W^{\mu\alpha\nu\beta} = \hat{W}^{\mu\alpha\nu\beta} - V^{\mu\alpha\nu\beta}.$$

The presence of this V-tensor in the UV divergences of QED obliges us to include a counter term in the photon action of the kind discussed above:

$$\propto C^{\mu\alpha\nu\beta}F_{\mu\alpha}(x)F_{\nu\beta}(x)$$

The remaining $\hat{W}$-term relates to the renormalisation of the metric and other parameters.

We have then the following picture. If Lorentz invariance is broken with a new metric for the electron field, the resulting structure for the UV divergences forces explicit Lorentz symmetry breaking on the photon field with consequential birefringence.

Actually there is an exception to this situation. The electron and photon metrics may both be invariant under a subgroup of the Lorentz group with an invariant 4-vector. The 4-vector may be time-like space-like or light-like [2, 9]. In these circumstances the V-tensor vanishes and there is no need to introduce a birefringence inducing term for the photon field. The lack of equality between the two metrics still gives rise to Lorentz symmetry breaking.

8. Conclusions

We examined simple models in which Lorentz symmetry was violated by the presence of different space-time metrics associated with different fields. In order to maintain a causal structure we showed in the case of two interacting scalar fields that the two light-cones associated with the two metrics must not only have in common a set of planes space-like in both but must also overlap in such a way as to share a set of vectors that are time-like in both. This makes possible the existence of Minkowski rotations applicable simultaneously to both metrics in a way consistent with causality.

We considered electrodynamics in a premetric formulation which is the most general model encoding Lorentz symmetry violation. By establishing a preferred metric it was possible to identify Lorentz symmetry violation with an associated Weyl-like tensor (WLT) and to use the Petrov classification to identify different types of violation and resulting birefringence. When the Lorentz violation is viewed as originating in a separate metric for the electron field we indicated how perturbative renormalisation in general requires the introduction of the type of counter-terms for the photon field corresponding to a WLT and hence birefringence. This can be avoided when the two metrics are invariant under an appropriate subgroup of the Lorentz group. There is still Lorentz symmetry violation but no birefringence.

Acknowledgments

I am grateful for partial support from STFC grant ST/L000385/1.

References

[1] Drummond I T 2013 *Phys. Rev.* D **88** 025009
[2] Coleman S and Glashow S L 1999 *Phys. Rev.* D **59** 116008
[3] Colladay D and Kostelecky V A 1997 *Phys. Rev.* D **55** 6760
[4] Colladay D and Kostelecky V A 1998 *Phys. Rev.* D **58** 116002
[5] Kostelecky V A 2011 *Rev. Mod. Phys.* **83** 11
[6] Drummond I T and Hathrell S J 1980 *Phys. Rev.* D **22** 343
[7] Hollowood T J and Shore G M 2008 *Nucl. Phys.* B **795** 138
[8] Itin Y 2009 *J. Phys. A: Math. Theor.* **42** 475402
[9] Drummond I T 2017 *Phys. Rev.* D **95** 025006
[10] Mewes M and Kostelecky V A 2009 *Phys. Rev.* D **80** 015020
[11] Petrov A Z 1954 *Uch. Zapiski Kazan Gos. Univ.* **144** 55
[12] Newman E T and Penrose R 1962 *J. Math. Phys.* **3** 566
[13] Penrose R and Rindler W 1986 *Spinors and spacetime 2: Spinor and twistor methods in spacetime geometry* (Cambridge University Press, Cambridge)
[14] Stewart J 1991 *Advanced general relativity* (Cambridge Monographs on Mathematical Physics, Cambridge University Press)
[15] O' Donnell P 2003 *Introduction to 2-spinors in general relativity* (World Scientific Publishing Co. Pte. Ltd.)
[16] Klinkhamer F R and Schreck M 2011 *Nucl. Phys.* B **848** 90

Hořava gravity: motivation and status

Diego Blas

Theoretical Physics Department, CERN, Geneva, Switzerland

Abstract. In this short contribution, I will review the status of Hořava gravity as a proposal for quantum gravity. After summarizing the framework and phenomenological constraints, I will present some recent work on the renormalization group flow in $2+1$ dimensions. The main result is that the theory is asymptotically free and flows to an infrared fixed point connected to a Lorentz-invariant theory.

1. Introduction

The proposal of P. Hořava [1] stems from a simple question: *is there a viable, unitary quantum field theory of gravity in* $3+1$ *dimensions described by a perturbative fixed point at high energies?* The answer to this question seems to be negative in Lorentz-invariant theories. Dropping the different requirements one finds a diversity of proposals for quantum gravity with various levels of complexity and success. As an example, one can consider *higher-derivative gravity* [2]. This yields a renormalizable theory of quantum gravity, which can even be asymptotically free in $3+1$ dimensions [3]. However, the theory has problems with perturbative unitarity and stability, both related to the presence of higher time derivatives. Inspired by this proposal, P. Hořava suggested to study gravitational theories with higher spatial derivatives, but only *two* time derivatives to preserve unitarity. Schematically, one writes for the free part of the graviton field h (we do not include indexes in this expression for presentation purposes)

$$S = \int \mathrm{d}^3x\mathrm{d}t \left(\dot{h}\dot{h} - h(-\Delta)^z h\right). \tag{1}$$

If this term is dominant at high energies, the proper scaling to understand the relevance of interactions is

$$t \to \lambda^{-z} t, \quad x^i \to \lambda^{-1} x^i, \quad h \to \lambda^{3-z} h. \tag{2}$$

As a consequence, for $z = 3$, h is dimensionless, and the operators with additional factors of h, with the number of spatial derivates bounded by $2z$ and only two time derivatives, are all marginal or relevant. If the number of such operators is finite, one deals with a power-counting renormalizable theory.

2. Hořava gravity in a nutshell

The scaling laws (2) and the action (1) break the boost transformations of Lorentz invariance. The implication when building a gravitational theory is that the invariance under diffeomorphisms must be broken to a subgroup compatible with (2). To do this, one assumes that space-time is endowed with a preferred time foliation, which selects a preferred time direction t at each point of space-time. This structure preserves the symmetries

$$x^i \mapsto \tilde{x}^i(x,t), \quad t \mapsto \tilde{t}(t), \tag{3}$$

called foliation-preserving diffeomorphism. A gauge theory invariant under (3) requires the introduction of a metric,

$$\mathrm{d}s^2 = N^2\mathrm{d}t^2 - \gamma_{ij}(\mathrm{d}x^i + N^i\mathrm{d}t)(\mathrm{d}x^j + N^j\mathrm{d}t), \tag{4}$$

with the subtlety that the field N can depend only on time (*projectable* case) or be generic (*non-projectable*). The gravitational dynamics of the theory follows from the action built from the metric, invariant under (3) and having a maximum of two time derivatives and six space derivatives (to realize the relevant scaling at high energies). Schematically,

$$S = \frac{\tilde{M}_P}{2}\int \mathrm{d}^3x\mathrm{d}t\sqrt{\gamma}N\left[K_{ij}K^{ij} - (1-\lambda')K^2 - \eta R - \alpha a_i a^i + \cdots + \frac{\Delta^2 R^2}{M_\star^4}\right], \tag{5}$$

where K_{ij} is the extrinsic curvature of the foliations, and $a_i \equiv \partial_i \log N$. The constant $\tilde{M}_P$ is a renormalization of the Planck mass M_P by the parameters in the theory [4]. The symbol Δ refers to the three-dimensional Laplacian $\gamma^{ij}\nabla_i\nabla_j$ and R is the three-dimensional Ricci scalar. In fact, here and in the following the quantities referring to intrinsic geometrical properties are those of the three-dimensional metric γ_{ij}. For instance, R_{ij} will be the three-dimensional Ricci tensor. The ellipses represent all the operators with more than two space derivatives. There is a finite number of them and we assume that all of them are suppressed by a common scale $M_\star$. The terms before the ellipses are enough to describe the theory at energies $E < M_\star$. For the proposal to accomplish its purpose, the new high-energy scaling law should be effective before the theory becomes strongly coupled, $M_\star < \sqrt{c}\,M_P$, where c is a combination of the low-energy parameters representing the 'largest' of them [5]. This form of Hořava gravity was first presented in [6].

The projectable case implies $a_i = 0$. In this case, the number of spatial operators is relatively small in $3+1$ dimensions. Denoting by $\mathcal{V}$ the part of (5) containing only space derivatives, one finds

$$\mathcal{V} = 2\Lambda - \eta R + \mu_1 R^2 + \mu_2 R_{ij}R^{ij} + \nu_1 R^3 + \nu_2 R R_{ij} R^{ij} + \nu_3 R^i_j R^j_k R^k_i + \nu_4 \nabla_i R \nabla^i R + \nu_5 (\nabla_i R_{jk})^2, \tag{6}$$

with arbitrary constants. Note that the constants μ_i and ν_i are dimensionful, and we assume that its order of magnitude is given a common scale $M_\star$. Unfortunately, the Minkowski background is unstable in the projectable case in $3+1$ dimensions. This can be seen by realizing that the scalar (s) and tensor (TT) perturbations around it obey dispersion relations

$$\omega_s^2 = \frac{\lambda}{3\lambda - 2}(-\eta k^2 + \mu_{12}k^4 + \nu_{45}k^6), \quad \omega_{TT}^2 = \eta k^2 + \mu_2 k^4 + \nu_5 k^6. \tag{7}$$

The different parameters in this formula are combinations of the constants in (6), cf. [7]. Since $\frac{\lambda}{3\lambda-2} > 0$ from the absence of ghosts, we see that there is no choice: either the scalar or the tensor modes are unstable. The non-projectable case does not suffer from this problem, neither from any other known instability. However, since $a_i \neq 0$, the number of operators is substantially larger than in the projectable case. Furthermore, for the projectable case, it has been shown that renormalizability happens beyond power counting, which is a non-trivial check for gauge theories [7, 8]. The similar proof for the non-projectable case is currently missing due to the difficulty in treating the instantaneous modes present in the theory.

3. Phenomenology of non-projectable Hořava gravity

Since no instability is known for the non-projectable case, it is worth studying further other consequences of the previous proposal for gravitational physics. For the rest of fields in the standard model of particle physics Lorentz invariance has been tested to high accuracy [9, 10, 11], while for dark matter these tests has been proposed in [12, 13].

The gravitational tests of the proposal have been summarized in [4]. They are related to the presence of new massless extra degrees of freedom, modification of the graviton's properties and other tests of the existence of a preferred frame. One can establish a first classification: tests of the physics at energies $E > M_\star$ or at $E < M_\star$. In the first case, one explores the core of the proposal: the theory deviates completely from general relativity. For instance, Newton's law will be modified by an expression of the form

$$\Delta\left(1 + a_1\frac{\Delta}{M_\star^2} + \dots\right) = \frac{\tau^{00}}{2M_P^2}, \tag{8}$$

where by τ^{00} I mean the time-time component of the energy momentum tensor. The new terms are important to resolve the problem of singularities in this proposal, though the study of the structure of singularities is at its infancy, see e.g. [14]. They also imply that the potential generated by a point-mass m_{pp} will be

$$\phi_{pp} = -\frac{m_{pp}}{8\pi M_P^2 r}F(rM_\star), \tag{9}$$

where $F[rM_\star] \to 1$ at $rM_\star \gg 1$. At smaller distances, there is no analytic solution, but it will be less singular than the Newtonian potential, even finite as $r \to 0$. One can compare this prediction to the standard tests of deviations from Newton's law, yielding an estimate[15, 16] $M_\star \gtrsim 1/\mu\mathrm{m}$. However, these studies derive the constraints for potentials which are not solutions of (8) and that do not modify the singular behaviour of the Newtonian potential. It would thus be interesting to re-analyze the available data to establish more realistic constraints on the scale $M_\star$.

At lower energies, the theory is characterized by three dimensionless parameters beyond general relativity: λ', α, η. By using solar system tests [5], observations from pulsars [17] and cosmology [12] these parameters can be constrained to high precision, from $O(10^{-7})$ to $O(10^{-2})$ depending on the combination of parameters. This is remarkable, since one could detect predictions of a quantum gravity proposal at low energies. It is important to state that in the constrained region *non-projectable Hořava gravity agrees with all data.* The level of precision of the tests has been recently boosted by the strong bound on the speed of propagation of gravitational waves [18, 19]. This means that a particular combination of the infrared parameters is smaller than $O(10^{-15})$. If all the parameters λ', α, η were so small, the chances to test the proposal with astrophysical observations would be severely reduced. This is not problematic for the proposal, since the only fundamental constraint is $\sqrt{c}M_P > M_\star$ (recall our discussion in the previous section).

4. Main open questions for Hořava gravity

Despite the large number of works about Hořava gravity, some of the key questions are still not answered. Remarkably, among them we find the most interesting (and hard) ones:

- *Is the theory renormalizable beyond power-counting?* As I explained, this has been proven for the *projectable* case, while for the non-projectable case the question still remains. The difference is the presence of instantaneous modes in the latter, whose effects are harder to track beyond linear theory [7].

- *Is the theory asymptotically free/scale invariant?* A positive answer to the question in the previous point can still hide a Landau pole. In the next section I report on the first answer to this question in $2+1$ dimensions.
- *How can one explain the high precision at which Lorentz invariance holds for the standard model if it is not a fundamental symmetry?* In a sense, this is the bone in the throat of many people working in Lorentz violation. We have traded a hard problem (quantum gravity in $3+1$ dimensions) by another hard problem: explaining the emergence of Lorentz invariance. The possibility of Lorentz invariance as an emergent phenomenon has been studied in the past [20]. Recently different ways have been suggested to try to explain the tight tests of Lorentz invariance at low energies [10]:
  - First, it is technically natural to have *all* the parameters that break Lorentz invariance small. Still, one would like to find ways to make those related to gravity larger (to be able to test the proposal at low energies). Furthermore, these parameters can not be smaller than $\sqrt{c}M_P > M_\star > 1/\mu\text{m}$.
  - The hierarchy between gravitational parameters and standard-model parameters can be due to the fact that these sectors are connected by Planck-suppressed operators [21]. This idea has been further developed in [22].
  - By using supersymmetry, Lorentz invariance can be understood as an accidental symmetry [23]. However, it seems hard to extend this idea to Hořava gravity [24].
  - Finally, Lorentz invariance can be emergent in the infrared along renormalization group flows [20, 10, 25]. This proposal is very interesting, though to explain the small departures from Lorentz invariance at low energies, while allowing a substantial deviation at high-energies, one needs to resort to strong dynamics. The study of the efficiency of this mechanism is still under development. In Hořava gravity, one can hope that the theory is asymptotically free, which would allow for a lattice formulation of the proposal where the problem of emergence through the renormalization group flow can be studied in a fully consistent way.
- Finally, as we discussed before, the nature of black holes in these theories is not clear. Singularities may be solved if the theory is asymptotically free. Also the concept of horizon is modified, since the theory has modes propagating at arbitrary high speeds [14, 26, 27].

I expect that there will be progress in all the previous points. In particular, the bounds on phenomenology are so tight that it is worth now turning back to more 'theoretical' aspects if we want to keep exploring the proposal. In the following I report on a recent result about the ultraviolet behaviour of the theory.

5. Renormalizability of projectable case in $2+1$ dimensions

In the program of studying the renormalizability of the $3+1$ non-projectable Hořava gravity, there are many obstacles that can be first addressed in simpler settings. In particular, making explicit loop calculations with gravitational degrees of freedom and without Lorentz invariance is technically challenging. This is why, as a warm up, we studied the $2+1$ projectable case in [8] (see also [28]). This toy model can also be important to understand Hořava gravity as a ultraviolet complete theory of quantum gravity. In particular, the behaviour of singularities, horizons, strong-coupling dynamics (including gravity) or implications to AdS/CFT may be studied first in this context.

Projectable Hořava gravity in $2+1$ has a very simple action in terms of three couplings,

$$S = \frac{1}{2G}\int \mathrm{d}x^2\mathrm{d}t\sqrt{\gamma}\left(K_{ij}K^{ij} - \lambda K^2 - \mu R^2\right). \tag{10}$$

The term proportional to the 2-dimensional Ricci scalar, R, is absent given its topological nature, while we set the cosmological constant to zero. Out of the previous couplings, only *two* combinations are physical due to gauge invariance [8]: $\{\mathcal{G} \equiv G/\sqrt{\mu}, \lambda\}$. The technical details on how to perform the one-loop calculation and hence obtain the flow of the parameters in (10) are given in [8]. The final result is shown in Fig. 1, taken from [8].

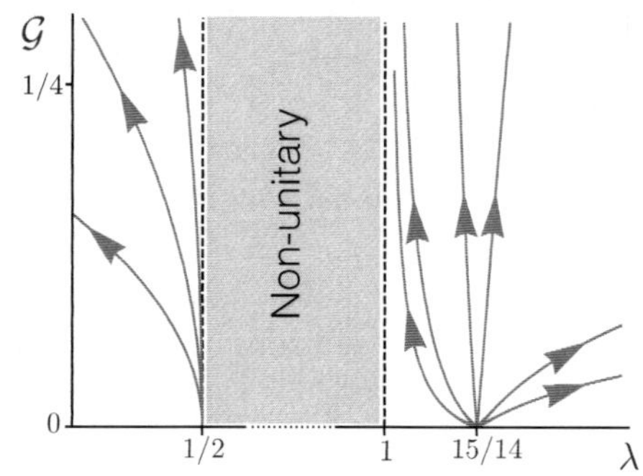

Figure 1. Renormalization group flow of the relevant couplings of eq. (10) at one loop. The arrows point towards the infrared. The region 'non-unitary' corresponds to a choice of parameters which has problems with unitarity.

There are several interesting aspects of the flow: first, the theory has two ultraviolet fixed points. The study of the $\lambda = 1/2$ fixed point requires a two-loop calculation that is beyond the purpose of [8]. More interestingly, the one at $\lambda = 15/14$ is asymptotically free, and the one-loop calculation is applicable. Thus, this theory is ultraviolet complete around this gaussian fixed point. Second, this fixed point approaches $\lambda \to 1$ in the infrared, which corresponds to the value of a Lorentz-invariant theory at low energies. We do not know if the theory is really close to $2+1$ general relativity since it becomes strongly coupled in this limit. This stems from the absence of the R term. This difficulty is not expected in $3+1$, where the new scaling generated by terms of the form R would stop the flow, and this can happen before the theory becomes strongly coupled.

Thus, one sees that the best possible scenario is satisfied in projectable $2+1$ Hořava gravity: asymptotically free theory flowing to a 'Lorentz invariant' fixed point in the infrared. Whether this will also be true in the realistic cases in $3+1$ dimensions is still to be studied.

6. Summary and outlook

In this contribution I have motivated Hořava gravity as a proposal for perturbative and unitary quantum gravity in $3+1$ dimensions. This is done at the expense of breaking of one of the better tested symmetries: boost invariance. Still, the parameters of the theory can be chosen to make the proposal compatible with all phenomenological tests.

There are many aspects of the theory that require further study. In my opinion, the emergence of Lorentz invariance at low energies, aspects of black holes (thermodynamics and singularities) and the computation of quantum properties are the most pressing ones. They are also the most challenging ones. Concerning the last point, I have reported on a fresh result about the one-loop calculation of the projectable case in $2+1$ dimensions, where we find a very positive outcome. The extension to the realistic $3+1$ non-projectable case or the non-perturbative study of the proposal requires a formidable effort. Still, our findings are encouraging enough to vigorously pursue this endeavour. Finally, it is necessary to include matter fields, to make sure that the nice properties we discussed are still preserved in a realistic scenario.

References

[1] Horava P 2009 *Phys. Rev.* D **79** 084008 (*Preprint* 0901.3775)
[2] Stelle K S 1977 *Phys. Rev.* D **16** 953–969
[3] Avramidi I G and Barvinsky A O 1985 *Phys. Lett.* **159B** 269–274
[4] Blas D and Lim E 2015 *Int. J. Mod. Phys.* D **23** 1443009 (*Preprint* 1412.4828)
[5] Blas D, Pujolas O and Sibiryakov S 2011 *J. High Energy Phys.* JHEP04(2011)018 (*Preprint* 1007.3503)
[6] Blas D, Pujolas O and Sibiryakov S 2010 *Phys. Rev. Lett.* **104** 181302 (*Preprint* 0909.3525)
[7] Barvinsky A O, Blas D, Herrero-Valea M, Sibiryakov S M and Steinwachs C F 2016 *Phys. Rev.* D **93** 064022 (*Preprint* 1512.02250)
[8] Barvinsky A O, Blas D, Herrero-Valea M, Sibiryakov S M and Steinwachs C F 2017 (*Preprint* 1705.03480)
[9] Coleman S R and Glashow S L 1999 *Phys. Rev.* D **59** 116008 (*Preprint* hep-ph/9812418)
[10] Liberati S 2013 *Class. Quant. Grav.* **30** 133001 (*Preprint* 1304.5795)
[11] Kostelecky V A and Russell N 2011 *Rev. Mod. Phys.* **83** 11–31 (*Preprint* 0801.0287)
[12] Audren B, Blas D, Ivanov M M, Lesgourgues J and Sibiryakov S 2015 *J. Cosmol. Astropart. Phys.* JCAP03(2015)016 (*Preprint* 1410.6514)
[13] Bettoni D, Nusser A, Blas D and Sibiryakov S 2017 *J. Cosmol. Astropart. Phys.* JCAP05(2017)024 (*Preprint* 1702.07726)
[14] Blas D and Sibiryakov S 2011 *Phys. Rev.* D **84** 124043 (*Preprint* 1110.2195)
[15] Adelberger E G, Heckel B R and Nelson A E 2003 *Ann. Rev. Nucl. Part. Sci.* **53** 77–121 (*Preprint* hep-ph/0307284)
[16] Kostelecky V A and Mewes M 2017 *Phys. Lett.* B **766** 137–143 (*Preprint* 1611.10313)
[17] Yagi K, Blas D, Yunes N and Barausse E 2014 *Phys. Rev. Lett.* **112** 161101 (*Preprint* 1307.6219)
[18] Blas D, Ivanov M M, Sawicki I and Sibiryakov S 2016 *JETP Lett.* **103** 624–626 [Pisma Zh. Eksp. Teor. Fiz.103,no.10,708(2016)] (*Preprint* 1602.04188)
[19] Abbott B P *et al.* (Virgo, Fermi-GBM, INTEGRAL, LIGO Scientific) 2017 *Astrophys. J.* **848** L13 (*Preprint* 1710.05834)
[20] Froggatt C and Nielsen H 1991 *Origin of Symmetries* (World Scientific) ISBN 9789971966317 URL https://books.google.ch/books?id=iSsM93vmsfgC
[21] Pospelov M and Shang Y 2012 *Phys. Rev.* D **85** 105001 (*Preprint* 1010.5249)
[22] Coates A, Colombo M, Gumrukcuoglu A E and Sotiriou T P 2016 *Phys. Rev.* D **94** 084014 (*Preprint* 1604.04215)
[23] Groot Nibbelink S and Pospelov M 2005 *Phys. Rev. Lett.* **94** 081601 (*Preprint* hep-ph/0404271)
[24] Marakulin A and Sibiryakov S 2016 (*Preprint* 1610.07805)
[25] Bednik G, Pujolas O and Sibiryakov S 2013 *J. High Energy Phys.* JHEP11(2013)064 (*Preprint* 1305.0011)
[26] Barausse E, Jacobson T and Sotiriou T P 2011 *Phys. Rev.* D **83** 124043 (*Preprint* 1104.2889)
[27] Liberati S and Pacilio C 2017 (*Preprint* 1709.05802)
[28] Griffin T, Grosvenor K T, Melby-Thompson C M and Yan Z 2017 *J. High Energy Phys.* JHEP06(2017)004 (*Preprint* 1701.08173)

Anomalous Lorentz and CPT violation

F R Klinkhamer

Institute for Theoretical Physics, Karlsruhe Institute of Technology (KIT), 76128 Karlsruhe, Germany

E-mail: `frans.klinkhamer@kit.edu`

Abstract. If there exists Lorentz and CPT violation in nature, then it is crucial to discover and understand the underlying mechanism. In this contribution, we discuss one such mechanism which relies on four-dimensional chiral gauge theories defined over a spacetime manifold with topology $\mathbb{R}^3 \times S^1$ and periodic spin structure for the compact dimension. It can be shown that the effective gauge-field action contains a local Chern-Simons-like term which violates Lorentz and CPT invariance. For arbitrary Abelian $U(1)$ gauge fields with trivial holonomies in the compact direction, this anomalous Lorentz and CPT violation has recently been established perturbatively with a Pauli–Villars-type regularization and nonperturbatively with a lattice regularization based on Ginsparg-Wilson fermions.

1. Introduction

Experiment has shown the violation of P, C, CP, and T, but *not* of CPT. Indeed, there is the well-known CPT "theorem" (Lüders, 1954–57; Pauli, 1955; Bell, 1955; Jost, 1957), which states that any local relativistic quantum field theory is invariant under the combined operation (in whichever order) of charge conjugation (C), parity reflection (P), and time reversal (T). The formulation of the theorem can be sharpened, but the form stated suffices for the moment.

The main inputs of the theorem are

- flat Minkowski spacetime $(M,\, g_{\mu\nu}) = \big(\mathbb{R}^4,\, \eta_{\mu\nu}^{\text{Minkowski}}\big)$;
- invariance under proper orthochronous Lorentz transformations and spacetime translations;
- standard spin-statistics connection;
- locality and Hermiticity of the Hamiltonian.

We refer to two monographs [1, 2] for further details and references on the CPT theorem.

The following question arises: *can* CPT invariance be violated at all in a physical theory and, if so, *is* it in the real world? It was widely believed that only quantum-gravity or superstring effects could give CPT violation. But a different result has been obtained several years ago: for certain spacetime topologies and classes of chiral gauge theories, CPT invariance is broken anomalously, that is, by quantum effects.

The original paper for this "CPT anomaly" is Ref. [3]. Follow-up papers have appeared in Refs. [4, 5, 6, 7, 8] and an extensive review is given by Ref. [9]. The crucial ingredients of the CPT anomaly are

- chiral fermions and gauge interactions;
- non-simply-connected spacetime topology,
- periodic boundary conditions.

IOP Conf. Series: Journal of Physics: Conf. Series **952** (2018) 012003 doi:10.1088/1742-6596/952/1/012003

Expanding on the last two bullets, the spacetime manifold must have a separable compact spatial dimension (coordinate x^3) with periodic spin structure of the fermions. The main focus, up till now, has been on the topology $\mathbb{R}^3 \times S^1$, but also other topologies have been considered, for example topologies related to punctures or wormholes [7].

There are, at least, three possible applications of the CPT anomaly. First, there is the resulting optical activity of the vacuum, which may lead to observable effects for the Cosmic Microwave Background [9, 10, 11]. Second, there is the resulting fundamental arrow-of-time, which may play a role in explaining the "start" of the Big Bang [12]. Third, the CPT anomaly may also be used as a diagnostic tool for a hypothetical spacetime foam [7, 13].

In this short contribution, we focus on the basic mechanism of the CPT anomaly and skip possible applications (referring to the review [9] for further discussion of the phenomenology).

2. Heuristics

The CPT anomaly of a chiral gauge theory defined over the four-dimensional manifold $M = \mathbb{R}^3 \times S^1$, with trivial vierbeins $e^a_\mu(x) = \delta^a_\mu$ and appropriate background gauge fields, arises in four steps:

- a compact spatial dimension with coordinate $x^3 \in [0, L]$ and a periodic spin structure make that a chiral fermion can have a momentum component $p_3 = 0$;
- a single chiral 2-component Weyl fermion in four dimensions (4D) with $p_3 = 0$ corresponds to a single massless 2-component Dirac fermion in three dimensions (3D);
- a single massless Dirac fermion in 3D is known to have a "parity anomaly," provided gauge invariance is maintained exactly [14, 15, 16];
- this "parity" violation in 3D corresponds to T violation in 4D, which, in turn, implies CPT violation in 4D.

The actual CPT anomaly will be established by calculating the effective gauge-field action $\Gamma[A]$, where the effects of the virtual fermions have been integrated out, and by showing that this effective action $\Gamma[A]$ changes under a CPT transformation of the background gauge field A.

3. Perturbative calculation

3.1. Main result

Consider, for definiteness, a chiral gauge theory with the following gauge group G, left-handed-fermion representation R_L, and spacetime manifold M:

$$G = SO(10)\,, \tag{1a}$$

$$R_L = N_{\text{fam}} \times (\mathbf{16})\,, \quad N_{\text{fam}} = 1\,, \tag{1b}$$

$$M = \mathbb{R}^3 \times S^1_{\text{PSS}}\,, \quad e^a_\mu(x) = \delta^a_\mu\,, \quad g_{\mu\nu}(x) = e^a_\mu(x)\, e^b_\nu(x)\, \eta_{ab} = \eta_{\mu\nu}\,, \tag{1c}$$

where the subscript "PSS" in (1c) stands for periodic spin structure and the metric $g_{\mu\nu}(x)$ has a Lorentzian signature $(-1, 1, 1, 1)$. The spacetime manifold M from (1c) is described by the following coordinates x^μ:

$$x^0, x^1, x^2 \in \mathbb{R}\,, \quad x^3 \in [0, L]\,. \tag{2}$$

The spinorial $SO(10)$ representation **16** from (1b) is a complex representation (see also the remarks in the first paragraph of Sec. 4.1). For simplicity, we have set N_{fam} in (1b) equal to 1, but similar results hold for the value $N_{\text{fam}} = 3$ relevant to elementary particle physics.

The fermionic field in the representation (1b) is denoted by $\psi_L(x)$ and the Lie-algebra-valued gauge field $A_\mu(x)$ is defined as follows:

$$A_\mu(x) \equiv g\, A_\mu^a(x)\, T^a\,, \tag{3a}$$

$$\mathrm{tr}\left(T^a\, T^b\right) = -\frac{1}{2}\,\delta^{ab}\,, \tag{3b}$$

with argument x standing for (x^0, x^1, x^2, x^3), Yang–Mills coupling constant g, and real gauge fields $A_\mu^a(x)$. The index "a" in (3a) labels the anti-Hermitian generators T^a of the so(10) Lie algebra and is summed over ($a = 1, \ldots, 45$), while (3b) specifies the normalization of the generators.

Both gauge and fermionic fields are assumed to have periodic boundary conditions in x^3,

$$A_\mu(\widetilde{x}, x^3 + L) = A_\mu(\widetilde{x}, x^3)\,, \tag{4a}$$

$$\psi_L(\widetilde{x}, x^3 + L) = \psi_L(\widetilde{x}, x^3)\,, \tag{4b}$$

with $\widetilde{x} \equiv (x^0, x^1, x^2)$. The coupling between gauge and fermionic fields is specified by the following Dirac equation (equivalent to the standard Weyl equation [17] by use of the chiral representation of the Dirac matrices γ^μ):

$$\gamma^\mu \left(\partial_\mu + A_\mu(x)\right) \psi_L(x) = 0\,, \tag{5}$$

where the coupling constant g has been absorbed in the gauge field according to (3a).

For this setup, the complete effective gauge-field action $\Gamma[A]$ for $A \in \mathrm{so}(10)$ is, of course, not known exactly (there are certain exact results in 2D [4, 5]). But the crucial term in $\Gamma[A]$ has been identified perturbatively [3, 8] for a gauge field with trivial holonomies (e.g., $A_3 = 0$):

$$\Gamma_{\mathrm{anom}}^{\mathbb{R}^3\times S^1}[A] = \frac{1}{16\,\pi^2} \int_{\mathbb{R}^3} dx^0 dx^1 dx^2 \int_0^L dx^3\, \frac{\pi}{L}\, \epsilon^{\kappa\lambda\mu 3}\, \mathrm{tr}\left(A_\kappa(x)\, A_{\lambda\mu}(x) - \frac{2}{3}\, A_\kappa(x)\, A_\lambda(x)\, A_\mu(x)\right) + \ldots\,, \tag{6}$$

in terms of the Levi-Civita symbol $\epsilon^{\kappa\lambda\mu\nu}$, the Lie-algebra-valued gauge field $A_\kappa(x)$ from (3a), and the corresponding Yang–Mills field strength $A_{\lambda\mu}(x)$,

$$A_{\lambda\mu}(x) \equiv \partial_\lambda A_\mu(x) - \partial_\mu A_\lambda(x) + A_\lambda(x)\, A_\mu(x) - A_\mu(x)\, A_\lambda(x)\,. \tag{7}$$

The ellipsis in (6) contains further (nonlocal) terms; see the third remark in Sec. 3.2.

The result (6) has been obtained perturbatively at the one-loop level for arbitrary gauge fields with trivial holonomies,

$$\forall \widetilde{x} \in \mathbb{R}^3 \;:\; H_3(\widetilde{x}) = \mathbb{1}\,, \tag{8}$$

where the holonomy along the straight path in the 3 direction starting at $\widetilde{x} \equiv (x^0, x^1, x^2)$ and $x^3 = 0$ is defined by

$$H_3(\widetilde{x}) \equiv \mathcal{P}\, \exp\left[\int_0^L dy\, A_3(\widetilde{x}, y)\right] \in G\,, \tag{9}$$

with the path-ordering operator $\mathcal{P}$. Taking into account the periodic boundary conditions (4a), the trace of the holonomy $H_3(\widetilde{x})$ corresponds to a gauge-invariant quantity, also known as the Wilson loop variable.

The spacetime integral in (6) involves the Chern-Simons density [18],

$$\omega_{\mathrm{CS}}[A_0, A_1, A_2] \equiv \frac{1}{16\,\pi^2}\, \epsilon^{klm}\, \mathrm{tr}\left(A_k\, A_{lm} - \frac{2}{3}\, A_k\, A_l\, A_m\right)\,, \tag{10}$$

where the repeated indices k, l, and m are summed over $\{0,\, 1,\, 2\}$. A genuine topological Chern–Simons term $\Omega_{\text{CS}} = \int \omega_{\text{CS}}$ is obtained only for a 3-dimensional spacetime manifold [18]. Our anomalous action term (6) holds, however, in four spacetime dimensions: the integration is over four spacetime coordinates and the gauge fields A_μ also have a dependence on x^3. Hence, the qualification "Chern–Simons-like" for the anomalous local action term in (6). This local action term is nontopological in the sense that there is a nontrivial dependence on the spacetime metric or vierbein; see, e.g., Ref. [19] and also the further remarks in the last paragraph of Sec. 5.

The local term (6) is *Lorentz-noninvariant*, because of the explicit spacetime index "3" entering the Levi–Civita symbol, and is also *CPT-odd*, because of the odd number (namely, three) of spacetime indices for the gauge-field terms in the large brackets. Recall that the standard Yang–Mills action density term tr $\big(A_{\mu\nu}(x)\, A^{\mu\nu}(x)\big)$ is Lorentz-invariant and CPT-even [17].

3.2. Technical remarks

In this subsection, we present six technical remarks, which can, however, be skipped in a first reading. First, the non-Abelian result (6) also holds for an Abelian $U(1)$ group, as long as attention is given to the proper normalization. Indeed, the single Lie algebra generator T^1 can be written as $T^1 = \big(i/\sqrt{2n}\,\big)\, \mathbb{1}_n$, in terms of an $n\times n$ identity matrix $\mathbb{1}_n$. Now take $n=1$, so that the trace operation in (6) becomes trivial. The covariant derivative (5) on a single left-handed fermion χ_L then reads $\big(\partial_\mu + i\,(g/\sqrt{2})\, A^1_\mu\big)\, \chi_L$ and we can define the Abelian $U(1)$ gauge field $B_\mu(x) \equiv A^1_\mu(x)$ and the $U(1)$ charge $e \equiv g/\sqrt{2}$. This rather explicit discussion aims to clarify some confusing statements in Ref. [3], where the parameter "a" in Eqs. (3.13) and (4.1) is simply to be omitted.

Second, the following expression is obtained by writing $\epsilon^{\kappa\lambda\mu 3}$ in the integrand of (6) as $\epsilon^{\kappa\lambda\mu\nu}\, \partial_\nu x^3$ and performing a partial integration (for gauge fields vanishing at infinity):

$$\Gamma^{\mathbb{R}^3\times S^1}_{\text{anom}}[\,A\,] = \frac{1}{32\,\pi} \int_{\mathbb{R}^3} dx^0 dx^1 dx^2 \int_0^L dx^3\, \frac{x^3}{L}\, \epsilon^{\kappa\lambda\mu\nu}\, \text{tr}\, \Big(A_{\kappa\lambda}(x)\, A_{\mu\nu}(x)\Big) + \ldots\,, \tag{11}$$

where the integrand depends solely on the Yang–Mills field strength (7). The spacetime-dependent "coupling constant" x^3/L in (11) makes clear that Lorentz invariance is broken.

Third, let us briefly discuss the issue of non-Abelian gauge invariance (further details can be found in Sec. 4 of Ref. [3]). The integrand of the non-Abelian Chern–Simons-like term in (6) is not completely gauge invariant (invariant only under infinitesimal 4D gauge transformations) but is manifestly periodic in x^3 because of the boundary conditions (4a). The integrand of the non-Abelian term (11) is completely gauge invariant (invariant also under "large" 3D gauge transformations and infinitesimal 4D deformations thereof) but is not manifestly periodic in x^3 (in addition, the single coordinate x^3 is not a "good" coordinate and two patches are needed to cover the circle S^1). Incidentally, the nonlocal terms of the ellipsis in (6) are precisely there to restore the full non-Abelian gauge invariance (cf. Refs. [15, 16]).

Fourth, the anomalous Chern–Simons-like term (6) for non-Abelian gauge fields has, strictly speaking, only been derived [3] for a special class of background fields, namely x^3-independent gauge fields with vanishing component in the compact direction,

$$A_3 \;=\; 0\,, \tag{12a}$$

$$A_{\widetilde{\mu}} \;=\; A_{\widetilde{\mu}}(\widetilde{x})\,, \tag{12b}$$

with definition $\widetilde{x} \equiv (x^0,\, x^1,\, x^2)$ and $\widetilde{\mu}$ taking values from the set $\{0,\, 1,\, 2\}$.

Fifth, it is possible to consider a larger class of background fields, namely x^3-dependent

IOP Conf. Series: Journal of Physics: Conf. Series **952** (2018) 012003 doi:10.1088/1742-6596/952/1/012003

Abelian $U(1)$ gauge fields $B_\mu(x) \in \mathbb{R}$ with vanishing component in the compact direction,

$$B_3 = 0\,, \tag{13a}$$
$$B_{\widetilde{\mu}} = B_{\widetilde{\mu}}(\widetilde{x},\, x^3)\,, \tag{13b}$$
$$B_{\widetilde{\mu}}(\widetilde{x},\, x^3) = B_{\widetilde{\mu}}(\widetilde{x},\, x^3 + L)\,, \tag{13c}$$

together with periodic boundary conditions (4b) on the relevant fermionic fields $\chi_L(x)$. For Abelian $U(1)$ gauge fields (13) with definitions from the first remark of this subsection and with a Pauli–Villars-type regularization [20], the Abelian version of the local Chern–Simons-like term in (6) has been obtained [8],

$$\Gamma^{\mathbb{R}^3\times S^1}_{\text{anom}}[B] = \frac{1}{16\,\pi^2}\,F\,e^2 \int_{\mathbb{R}^3} dx^0 dx^1 dx^2 \int_0^L dx^3\, \frac{2\,\pi}{L}\, \epsilon^{\kappa\lambda\mu 3}\, B_\kappa(x)\, \partial_\lambda B_\mu(x) + \ldots\,. \tag{14}$$

The numerical factor F in (14) counts the sum of the squares of the fermion charge in units of e; see also Sec. 4.2. The derivation of (14) requires that the typical momenta of the background fields (13) are very much less than the regulator masses.

Sixth, the anomalous term (6) can be generalized by including an additional prefactor $\big(2\,k^{(0)}+1\big)$ for $k^{(0)} \in \mathbb{Z}$, due to the freedom in defining the regularized theory. This freedom is particularly clear in the lattice formulation to be discussed in the next section, where the fermion measure of the path integral is characterized by an integer $k^{(0)}$.

4. Nonperturbative calculation

4.1. Main result

Consider, for definiteness, a four-dimensional Abelian chiral gauge theory with

$$G = U(1), \tag{15a}$$
$$R_L = 6 \times \left(\frac{1}{3}\right) + 3 \times \left(-\frac{4}{3}\right) + 3 \times \left(\frac{2}{3}\right) + 2 \times \big(-1\big) + 1 \times \big(2\big) + 1 \times \big(0\big)\,. \tag{15b}$$

The perturbative chiral gauge anomalies of this particular theory cancel out, because the group (15a) is a subgroup of the "safe" group $SO(10)$ and the reducible representation (15b) of $U(1)$ is contained in the irreducible representation **16** of $SO(10)$. Observe that R_L from (15b) is a complex representation: there are, for example, six left-handed fermions with a normalized charge $+1/3$ but none with a normalized charge $-1/3$. This may be compared to the vectorlike gauge theory of quantum electrodynamics with the real representation $R_L = (+1) + (-1)$.

Next, define a chiral lattice gauge theory over a finite hypercubic lattice with

- periodic spin structure in one direction;
- Ginsparg–Wilson fermions [21];
- Neuberger's lattice Dirac operator [22, 23];
- Lüscher's chiral constraints [24, 25].

The fermionic fields are, as usual, associated with the lattice sites and the gauge fields with the directed links between neighbouring lattice sites. Specifically, the fields are denoted by $\psi_L(x)$ at the lattice site x and $U_\mu(x) \in G$ for the link in the μ direction starting from the lattice site x.

The Euclidean effective gauge-field action $\Gamma[U]$ is given by the Euclidean path integral over the fermionic fields $\psi_L(x)$. The goal is to establish that this effective gauge-field action $\Gamma[U]$ changes under a CPT transformation,

$$\Gamma[U] \neq \Gamma[U^{\text{CPT}}]\,, \tag{16}$$

where U denotes the set of link variables of the lattice.

The result (16) has been established nonperturbatively [6, 8] for Abelian $U(1)$ gauge fields with trivial holonomies (e.g., $U_3 = 1$). Moreover, the origin of the CPT anomaly has been identified as an ambiguity [6] in the choice of basis vectors for the fermion integration measure; cf. the path-integral derivation of the triangle (Adler–Bell–Jackiw) anomaly [26].

The result (16) has, in particular, been obtained [8] for $U(1)$ link variables that are x^3-dependent and trivial in the compact direction,

$$U_3 = 1\,, \tag{17a}$$

$$U_{\widetilde{\mu}} = U_{\widetilde{\mu}}(\widetilde{x},\, x^3)\,, \tag{17b}$$

$$U_{\widetilde{\mu}}(\widetilde{x},\, x^3) = U_{\widetilde{\mu}}(\widetilde{x},\, x^3 + L)\,, \tag{17c}$$

with definition $\widetilde{x} \equiv (x^0,\, x^1,\, x^2)$ and $\widetilde{\mu}$ taking values from the set $\{0,\, 1,\, 2\}$. All arguments in (17) are understood to coincide with the points of the four-dimensional hypercubic lattice.

4.2. Continuum limit

In addition to establishing the existence of the CPT anomaly nonperturbatively, we have also investigated the continuum limit, with lattice spacing $a \to 0$ and number of links in the 3 direction $N_3 \to \infty$, while keeping $L \equiv N_3\, a$ at a constant value. For slowly-varying link variables $U_\mu(x) \approx \exp[i\, e\, a\, B_\mu(x)]$ in the continuum limit, the anomalous term $\Gamma_{\text{anom}}[U]$ from the nonperturbative lattice calculation has been shown [8] to give rise to the Abelian Chern–Simons-like term (14) with a numerical factor $F \equiv \sum_f (q_f/e)^2 = 40/3$ for the representation (15b). As mentioned in the last remark of Sec. 3.2, there is an additional prefactor $\left(2\, k^{(0)} + 1\right)$ coming from the definition of the path-integral fermion measure characterized by an integer $k^{(0)}$ (see App. B of Ref. [6]).

Hence, the nonperturbative lattice result does not modify the result of the perturbative one-loop calculation as discussed in Sec. 3. This is certainly reminiscent of the Adler–Bardeen result for the triangle anomaly [27], but it remains to be confirmed that there arise no additional terms in the nonperturbative lattice calculation.

5. Outlook

The subtle role of topology for the local properties of quantum field theory is well-known, the prime example being the Casimir effect. For certain chiral gauge theories, we now have established that the interplay of ultraviolet and infrared effects may also lead to Lorentz and CPT noninvariance, even for flat spacetime manifolds, that is, without local gravity.

Let us end with two general remarks. First, we have found an anomalous origin of Lorentz and CPT violation, which ultimately traces back to the regularization in the ultraviolet, operating at ultrahigh energies or ultra-small distances (see, e.g., the discussion in Sec. V of Ref. [8]). Regularization is typically used as a mere mathematical device, to be removed after the calculation. But it is also possible that a particular regularization has a deeper physical meaning. An example is the lattice regularization, which could be directly relevant if spacetime possesses a fundamental discreteness.

Second, the question arises how the anomalous photon term (14) affects gravity [28, 29]. The problem, now, is that the corresponding term in the energy-momentum tensor is, in general, neither symmetric nor conserved. There are solutions of the Einstein equation for special cases [29] but not for the general case. At this moment, it is not clear how the gravitational theory can be extended in order to allow for an anomalous Chern–Simons-like-term in the effective gauge-field action.

Acknowledgments

The author thanks M. Schreck for agreeing, on short notice, to present this talk at the workshop.

References

[1] Sakurai J J 1964 *Invariance Principles and Elementary Particles* (Princeton: Princeton University Press)
[2] Streater R and Wightman A 1964 *PCT, Spin and Statistics, and All That* (New York: Benjamin)
[3] Klinkhamer F R 2000 A CPT anomaly *Nucl. Phys.* B **578** 277 (preprint arXiv:hep-th/9912169)
[4] Klinkhamer F R and Nishimura J 2001 CPT anomaly in two-dimensional chiral $U(1)$ gauge theories *Phys. Rev.* D **63** 097701 (preprint arXiv:hep-th/0006154)
[5] Klinkhamer F R and Mayer C 2001 Torsion and CPT anomaly in two-dimensional chiral U(1) gauge theory *Nucl. Phys.* B **616** 215 (preprint arXiv:hep-th/0105310)
[6] Klinkhamer F R and Schimmel J 2002 CPT anomaly: A rigorous result in four dimensions *Nucl. Phys.* B **639** 241 (preprint arXiv:hep-th/0205038)
[7] Klinkhamer F R and Rupp C 2004 Space-time foam, CPT anomaly, and photon propagation *Phys. Rev.* D **70** 045020 (preprint arXiv:hep-th/0312032)
[8] Ghosh K J B and Klinkhamer F R 2018 Anomalous Lorentz and CPT violation from a local Chern–Simons-like term in the effective gauge-field action *Nucl. Phys.* B **926** 335 (preprint arXiv:1706.07025)
[9] Klinkhamer F R 2005 Nontrivial spacetime topology, CPT violation, and photons *Preprint* arXiv:hep-ph/0511030
[10] Carroll S M, Field G B and Jackiw R Limits on a Lorentz- and parity-violating modification of electrodynamics *Phys. Rev.* D **41** 1231 (1990)
[11] Komatsu E *et al.* [WMAP Collaboration] 2009 Five-year Wilkinson Microwave Anisotropy Probe (WMAP) observations: Cosmological interpretation *Astrophys. J. Suppl.* **180** 330 (preprint arXiv:0803.0547)
[12] Klinkhamer F R 2002 Fundamental time asymmetry from nontrivial space topology *Phys. Rev.* D **66** 047701 (preprint arXiv:gr-qc/0111090)
[13] Klinkhamer F R and Rupp C 2005 Photon-propagation model with random background field: Length scales and Cherenkov limits *Phys. Rev.* D **72** 017901 (preprint arXiv:hep-ph/0506071)
[14] Redlich A N 1984 Parity violation and gauge noninvariance of the effective gauge field action in three dimensions *Phys. Rev.* D **29** 2366
[15] Alvarez-Gaumé L, Della Pietra S and Moore G W 1985 Anomalies and odd dimensions *Annals Phys. (N.Y.)* **163** 288
[16] Coste A and Lüscher M 1989 Parity anomaly and fermion-boson transmutation in three-dimensional lattice QED *Nucl. Phys.* B **323** 631
[17] Itzykson C and Zuber J B 1980 *Quantum Field Theory* (New York: McGraw–Hill)
[18] Chern S S and Simons J 1974 Characteristic forms and geometric invariants *Annals Math.* **99** 48
[19] Kant E and Klinkhamer F R 2005 Maxwell–Chern–Simons theory for curved spacetime backgrounds *Nucl. Phys.* B **731** 125; 2009 Erratum *Nucl. Phys.* B **809** 362 (preprint arXiv:hep-th/0507162)
[20] Frolov S A and Slavnov A A 1993 An invariant regularization of the Standard Model *Phys. Lett.* B **309** 344
[21] Ginsparg P H and Wilson K G 1982 A remnant of chiral symmetry on the lattice *Phys. Rev.* D **25** 2649
[22] Neuberger H 1998 Exactly massless quarks on the lattice *Phys. Lett.* B **417** 141 (preprint arXiv:hep-lat/9707022)
[23] Neuberger H 1998 More about exactly massless quarks on the lattice *Phys. Lett.* B **427** 353 (preprint arXiv:hep-lat/9801031)
[24] Lüscher M 1998 Exact chiral symmetry on the lattice and the Ginsparg–Wilson relation *Phys. Lett.* B **428** 342 (preprint arXiv:hep-lat/9802011)
[25] Lüscher M 1999 Abelian chiral gauge theories on the lattice with exact gauge invariance *Nucl. Phys.* B **549** 295 (preprint arXiv:hep-lat/9811032)
[26] Fujikawa K 1980 Path integral for gauge theories with fermions *Phys. Rev.* D **21** 2848; 1980 Erratum: *Phys. Rev.* D **22** 1499
[27] Adler S L and Bardeen W A 1969 Absence of higher-order corrections in the anomalous axial-vector divergence equation *Phys. Rev.* **182** 1517
[28] Kostelecky V A 2004 Gravity, Lorentz violation, and the standard model *Phys. Rev.* D **69** 105009 (preprint arXiv:hep-th/0312310)
[29] Kant E 2009 Untersuchungen zu Lorentz-Verletzung, Gravitation und Photon-Propagation *PhD Thesis Universität Karlsruhe (TH)* [`http://inspirehep.net/record/1387559/files/Diss_Kant.pdf`]

IOP Conf. Series: Journal of Physics: Conf. Series **952** (2018) 012004 doi:10.1088/1742-6596/952/1/012004

Gravity, Lorentz violation, and effective field theory

J D Tasson

Physics and Astronomy Department, Carleton College, 1 North College St., Northfield, MN, 55057, USA

E-mail: jtasson@carleton.edu

Abstract. This proceedings contribution summarizes the effective field theory approach to searches for violations of Lorentz invariance in gravity with a focus on propagation effects.

1. Introduction

The question of how General Relativity (GR) and the Standard Model (SM) of particle physics merge to form a single consistent theory is one of the most fundamental questions we currently face in physics. Over the past several decades, the realization that certain scenarios for this merge could generate violations of local Lorentz invariance (Lorentz violation) [1] has led to the development of a framework known as the gravitational Standard-Model Extension (SME) [2, 3, 4], which is designed to search for Lorentz violation comprehensively and systematically. The SME is an effective field-theory framework that characterizes arbitrary Lorentz-violating corrections to GR and the SM. These corrections can be sought in existing experiments in hopes of gaining insight about the nature of Planck-scale physics with existing technology, and dozens of SME-based experiments have been performed to date [5].

The SME consists of an expansion about the action of GR and the SM in operators of increasing mass dimension d. The idea is to search systematically for Lorentz violation throughout physics rather than individually testing a large number of special models (though specific models [6] also provide useful insight). The approach resonates philosophically with other familiar test frameworks including the Parameterized Post Newtonian (PPN) formalism [7], while differing in its goals, construction, and physical effects. In terms of goals, the SME parameterizes deviations from exact Lorentz symmetry throughout physics, some of which are modifications to GR, while the PPN parameterizes differences from GR, some of which are Lorentz violating. By way of construction, the SME parameterizes deviations from the action of GR and the SM, while the PPN parameterizes deviations from the post-newtonian metric of GR. Another approach that has some philosophical similarity with the SME is the Parameterized Post Einsteinian (PPE) framework [8]. The key points are that (i) the SME is a broad and general test frame work having philosophical similarities with other test frameworks and (ii) the SME is unique in its comprehensive focus on Lorentz symmetry via an effective field-theory approach. While a number of approaches to Lorentz violation restrict attention to models that are isotropic in a special frame, the SME treats general anisotropic effects.

2. Linearized gravity

Much work has been done across a wide variety of gravitational systems in the SME. Lorentz violation in the pure-gravity sector at mass dimension four has been sought in a large number of systems including laboratory [9, 10, 11], solar-system [12, 13, 14], and astrophysical [15, 16, 17] tests. Much of this experimental and observational work is based on the theoretical and phenomenological developments of Ref. [18]. Searches for Lorentz violation in matter-gravity couplings have also been done [9, 11, 14, 19], initiated by the theoretical and phenomenological work of Refs. [20]. Studies of gravitational Lorentz violation beyond $d = 4$ have also begun. Here it has been shown that laboratory tests and gravitational-wave tests provide complementary coverage of the coefficient space [21]. Short-range gravity tests have made significant progress in this area [22], as have gravitational-wave tests [15, 17, 23, 24]. Exploration of the role of solar-system tests in searches for higher mass dimension operators has has also begun [25], as have considerations of nonlinear gravity effects [26]. We also note in passing that Finsler geometry provides a geometric interpretation for a number of Lorentz-violating effects in the effective field-theory construction [27].

In this section we focus on the full Lorentz-violating expansion about linearized gravity and the associated effects on the propagation of gravitational waves. Here the generic form of the Lagrange density including both the Lorentz-violating and Lorentz-invariant contributions can be written [23],

$$\mathcal{L}_{\mathcal{K}^{(d)}} = \tfrac{1}{4} h_{\mu\nu} \widehat{\mathcal{K}}^{(d)\mu\nu\rho\sigma} h_{\rho\sigma}, \tag{1}$$

where the operator

$$\widehat{\mathcal{K}}^{(d)\mu\nu\rho\sigma} = \mathcal{K}^{(d)\mu\nu\rho\sigma\varepsilon_1\varepsilon_2\ldots\varepsilon_{d-2}} \partial_{\varepsilon_1}\partial_{\varepsilon_2}\ldots\partial_{\varepsilon_{d-2}} \tag{2}$$

has mass dimension $d \geq 2$. Here $h_{\mu\nu}$ is the usual metric perturbation and the coefficients $\mathcal{K}^{(d)\mu\nu\rho\sigma\varepsilon_1\varepsilon_2\ldots\varepsilon_{d-2}}$ are taken as small constants. Intuition for Eq. (1) can be gained by noting that it contains the standard linearized limit of the Einstein–Hilbert action, which is recovered in the limit

$$\widehat{\mathcal{K}}^{(d)\mu\nu\rho\sigma} \to \epsilon^{\mu\rho\alpha\kappa}\epsilon^{\nu\sigma\beta\lambda}\eta_{\kappa\lambda}\partial_\alpha\partial_\beta. \tag{3}$$

In other cases, the coefficients $\mathcal{K}^{(d)\mu\nu\rho\sigma\varepsilon_1\varepsilon_2\ldots\varepsilon_{d-2}}$ parameterize the amount of Lorentz violation in the theory. A decomposition of the operator into irreducible parts generates 14 classes of operators. Three of these classes respect the usual gauge invariance of GR. These can be understood as being associated with spontaneous Lorentz violation [3, 28]. Initial exploration of these classes has been done [23].

Similar considerations in the photon sector lead to effects in vacuum that are analogous to the propagation of light in matter including energy-dependent birefringence, dispersion, and anisotropy. These effects have been sought astrophysically using a variety of observations including gamma-ray bursts [29, 30, 31] and CMB observations [29, 32]. The three classes of operators identified above are associated with the gravitational versions of the photon effects noted. These effects can be characterized via the dispersion relation

$$\omega = \left(1 - \varsigma^0 \pm \sqrt{(\varsigma^1)^2 + (\varsigma^2)^2 + (\varsigma^3)^2}\right)|\boldsymbol{p}|, \tag{4}$$

found from the relevant limit of (1) [23], where

$$\begin{aligned} \varsigma^0 &= \sum_{djm} \omega^{d-4}\, Y_{jm}(\hat{\boldsymbol{n}})\, k^{(d)}_{(I)jm}, \\ \varsigma^1 \mp i\varsigma^2 &= \sum_{djm} \omega^{d-4}\, {}_{\pm4}Y_{jm}(\hat{\boldsymbol{n}})\, \big(k^{(d)}_{(E)jm} \pm i k^{(d)}_{(B)jm}\big), \\ \varsigma^3 &= \sum_{djm} \omega^{d-4}\, Y_{jm}(\hat{\boldsymbol{n}})\, k^{(d)}_{(V)jm}. \end{aligned} \tag{5}$$

Here the relevant contributions to the coefficients $\mathcal{K}^{(d)\mu\nu\rho\sigma\varepsilon_1\varepsilon_2\ldots\varepsilon_{d-2}}$ can be decomposed into the spherical coefficients $k^{(d)}_{(I)jm}, k^{(d)}_{(E)jm}, k^{(d)}_{(B)jm}, k^{(d)}_{(V)jm}$ using spherical harmonics Y_{jm}, and spin-weighted spherical harmonics ${}_sY_{jm}$ [29]. The features in the analogy with optics in matter can be seen directly from (4). Birefringence is associated with the plus/minus in (4), dispersion with the appearance of factors of ω at each mass dimension $d > 4$, and anisotropy with the direction of origin for the wave $\hat{\boldsymbol{n}}$ via the spherical harmonics with $j \neq 0$.

3. Gravitational-wave applications

The first direct observations of gravitational waves from binary black-hole mergers were used to place initial direct constraints on modifications to the gravitational wave dispersion relation in the context of effective field theory. This was done through the absence of birefringence [23] in the first gravitational-wave observation [33] and the absence of dispersion [24] in the first [33] and second [34] gravitational-wave observations. These first analyses were performed via the chirp signals published by LIGO. The dispersive effects of Lorentz violation have subsequently been sought by the LIGO collaboration in the context of an isotropic phenomenological model [35]. Gravitational waves were also used indirectly in early work placing limits via the lack of gravitational Čerenkov radiation by scalar cosmic rays [36]. More recently, this idea has been expanded upon within the effective field-theory framework of the SME to include the full space of anisotropic effects, Lorentz violation associated with higher mass dimension operators, and gravitational Čerenkov radiation by other species including photons [15].

As an example of recent progress, the seminal observation of gravitational waves from a binary neutron-star merger by the LIGO-Virgo collaboration coincident with the observation of gamma-ray photons by the Fermi telescope provided an impressive new test of the effects contained in the gravitational dispersion relation [17]. The fact that the gamma-ray photons arrived about 1.7 seconds after the gravitational waves associated with the merger implies constraints on the relative speed of the gravitational waves and the photons [17] that are a significant improvement over prior direct speed-of-gravity tests [38]. Within effective field theory, the relative group velocity of the gravitational waves and the electromagnetic waves, $\Delta v = v_g - v_\gamma$, is controlled by differences in coefficients for Lorentz violation in the gravitational sector and the photon sector at each mass dimension. The explicit form of the group-velocity difference can be obtained from the dispersion relation in each sector [23, 29, 31] via standard methods. As birefringent coefficients can often be better sought with higher sensitivity using birefringence directly [23], we focus here on non-birefringent effects. Following this specialization, the resulting difference in group velocities for the two sectors can be written

$$\Delta v = -\sum_{djm}(d-3)Y_{jm}(\hat{\boldsymbol{n}})\left(E_g^{d-4}k^{(d)}_{(I)jm} - E_\gamma^{d-4}c^{(d)}_{(I)jm}\right). \tag{6}$$

Here E_g is the energy corresponding to the observed gravitational-wave frequency, and E_γ is the energy of the observed gamma rays. The result is presented in a spherical harmonic, Y_{jm}, basis, with $k^{(d)}_{(I)jm}$ and $c^{(d)}_{(I)jm}$ being the nonbirefingent spherical-basis coefficients for Lorentz violation in the gravity sector and electromagnetic sector, respectively. The sum is over even $d \geq 4$, and $j \leq d-2$. The direction $\hat{\boldsymbol{n}}$ refers to the direction of the incoming messengers from the event and is described by the standard spherical polar coordinates θ, ϕ in the Sun-centered celestial-equatorial frame [37]. Use of this frame is standard in SME studies as it facilitates easy comparison among results of different experiments.

The focus of the analysis performed by the LIGO, Virgo, Fermi Gamma-ray Burst Monitor, and INTEGRAL collaborations was on the minimal Lorentz-violating effects associated with operators of mass dimension $d = 4$, which are nondispersive and hence inaccessible to other gravitational wave-based approaches. Here, Eq. (6) can be written

$$\Delta v = -\sum_{jm} Y_{jm}(\hat{\boldsymbol{n}}) \left(\tfrac{1}{2}(-1)^{1+j}\overline{s}^{(4)}_{jm} - c^{(d)}_{(I)jm} \right), \tag{7}$$

where $k^{(4)}_{(I)jm} = \frac{1}{2}(-1)^{1+j}\overline{s}^{(4)}_{jm}$ [21] is used to facilitate contact with earlier SME work at $d = 4$. The resulting speed constraints generated a ten order of magnitude improvement over existing sensitivities to $\overline{s}^{(4)}_{00}$ [16, 12] and order of magnitude improvements in a number of other $d = 4$ gravity-sector coefficients in the context of a maximum-reach analysis [9] in which each of the gravity-sector coefficients were considered one at a time.

Before ending discussion of this result, we provide a few remarks to help the reader develop an intuition for spherical coefficients. First, note that based on the range of the indices in Eq. (6), nine $d = 4$ spherical coefficients $\overline{s}^{(4)}_{jm}$ contribute. This is consistent with the counting in the Cartesian basis where $\overline{s}^{\mu\nu}$ is symmetric and traceless. Note also that the coefficients satisfy $(\overline{s}^{(4)}_{jm})^* = (-1)^m \overline{s}^{(4)}_{j(-m)}$, where * is complex conjugation. The phase convention for the spherical harmonics is chosen such that they satisfy the same condition. Contact can then be made between the Cartesian and spherical coefficients by matching terms from each side of an expression such as

$$\overline{s}^{\mu\nu}\hat{p}_\mu \hat{p}_\nu = \sum_{jm} Y_{jm}(\hat{p})\overline{s}^{(4)}_{jm}. \tag{8}$$

For example, one finds $2\overline{s}^{YZ} = \sqrt{\frac{15}{2\pi}}\mathrm{Im}\overline{s}^{(4)}_{21}$. Reference [39] provides explicit relations for all nine coefficients. For additional details, see Ref. [29].

4. Summary

Effective field theory methods provide a framework for searching for Lorentz violation. This proceedings contribution summarizes this approach in the context of linearized gravity with a focus on propagation effects. As an example of the approach, we highlight the recent progress via multimessenger-astronomy observations. The future is bright for such searches as additional gravitational-wave observations distributed across the sky can be expected to generate tests that are increasingly sensitive and robust.

References

[1] Kostelecký V A and Samuel S 1989 *Phys. Rev. D* **39** 683
Kostelecký V A and Potting R 1991 *Nucl. Phys. B* **359** 545
[2] Colladay D and Kostelecký V A 1997 *Phys. Rev. D* **55** 6760
1998 *Phys. Rev. D* **58** 116002
[3] Kostelecký V A 2004 *Phys. Rev. D* **69** 105009
[4] For a review of the SME approach, see, Tasson J D 2014 Rep. Prog. Phys. **77** 062901
[5] *Data Tables for Lorentz and CPT Violation,* 2017 edition, Kostelecký V A and Russell N *Preprint* arXiv:0801.0287v10
[6] For models of local Lorentz violation, see for example, Jacobson T *Preprint* arXiv:0801.1547
Bluhm R *et al.* 2008 *Phys. Rev. D* **77** 065020
Altschul B *et al.* 2010 *Phys. Rev. D* **81** 065028
[7] For a review, see Will C M 1993 *Theory and Experiment in Gravitational Physics* (Cambridge, England: Cambridge University Press)
[8] Yunes N and Pretorius F 2009 *Phys. Rev. D* **80** 122003
[9] Flowers N A, Goodge C and Tasson J D accepted in *Phys. Rev. Lett. Preprint* arXiv:1610.05357
[10] Müller H, Chiow S, Herrmann S, Chu S and Chung K-Y 2008 *Phys. Rev. Lett.* **100** 031101
Chung K-Y, Chiow S, Herrmann S, Chu S and Müller H 2009 *Phys. Rev. D* **80** 016002
Bennett D, Skavysh V and Long J 2010 Search for Lorentz Violation in a Short-Range Gravity Experiment *CPT and Lorentz Symmetry V* ed Kostelecký V A (Singapore: World Scientific)

Bailey Q G, Everett R D and Overduin J M 2013 *Phys. Rev. D* **88** 102001
Müller H *et al.* 2008 *Phys. Rev. Lett.* **100** 031101
Shao C-G *et al. Preprint* arXiv:1707.02318
[11] Tasson J D 2012 *Phys. Rev. D* **86** 124021
Bourgoin A *et al. Preprint* arXiv:1706.06294
[12] Le Poncin-Lafitte C, Hees A and Lambert S 2016 *Phys. Rev. D* **94** 125030
[13] Battat J B R, Chandler J F and Stubbs C W 2007 *Phys. Rev. Lett.* **99** 241103
Bourgoin A *et al.* 2016 *Phys. Rev. Lett.* **117** 241301
Iorio L 2012 *Class. Quant. Grav.* **29** 175007
[14] Hees A *et al.* 2015 *Phys. Rev. D* **92** 064049
[15] Kostelecký V A and Tasson J D 2015 *Phys. Lett. B* **749** 551
Tasson J D 2016 *Symmetry* **8** 111
[16] Shao L 2014 *Phys. Rev. Lett.* **112** 111103
2014 *Phys. Rev. D* **90** 122009
[17] LIGO, Virgo, Fermi GBM and INTEGRAL Collaborations, Abbot B P *et al.* 2017 *Ap. J. Lett.* **848** L13
[18] Bailey Q G and Kostelecký V A 2006 *Phys. Rev. D* **74** 045001
[19] Hohensee M A, Müller H and Wiringa R B 2013 *Phys. Rev. Lett.* **111** 151102
Hohensee M A *et al.* 2011 *Phys. Rev. Lett.* **106** 151102
[20] Kostelecký V A and Tasson J D 2011 *Phys. Rev. D* **83** 016013
2009 *Phys. Rev. Lett.* **102** 010402
[21] Kostelecký V A and Mewes M 2017 *Phys. Lett. B* **766** 137
[22] Shao C-G *et al.* 2016 *Phys. Rev. Lett.* **117** 071102
Shao C-G *et al.* 2015 *Phys. Rev. D* **91** 102007
Long J and Kostelecký V A 2015 *Phys. Rev. D* **91** 092003
Bailey Q G, Kostelecký V A and Xu R 2015 *Phys. Rev. D* **91** 022006
[23] Kostelecký V A and Mewes M 2015 *Phys. Lett. B* **757** 510
[24] Yunes N, Yagi K and Pretorius F 2016 *Phys. Rev. D* **94** 084002
[25] Bailey Q G and Havert D 2017 *Phys. Rev. D* **96** 064035
[26] Bailey Q G 2016 *Phys. Rev. D* **94** 065029
[27] Kostelecký V A 2011 *Phys. Lett. B* **701** 137
Kostelecký V A, Russell N and Tso R 2012 *Phys. Lett. B* **716** 470
Silva J E G *Preprint* arXiv:1602.07345
Schreck M 2016 *Phys. Rev. D* **94** 025019
Foster J and Lehnert R 2015 *Phys. Lett. B* **746** 164
Aazami A B and Javaloyes M A 2016 *Class. Quant. Grav.* **33** 025003
Li X and Wang S 2016 *Eur. Phys J. C* **76** 51
Caponio E and Stancarone G 2016 *Int. J. Geom. Meth. Mod. Phys.* **13** 1650040
Minguzzi E 2016 *Rept. Math. Phys.* **77** 45
Russell N 2015 *Phys. Rev. D* **91** 045008
Colladay D and McDonald P 2015 *Phys. Rev. D* **92** 085031
[28] Bluhm R 2015 *Phys. Rev. D* **91** 065034
[29] Kostelecký V A and Mewes M 2009 *Phys. Rev. D* **80** 015020
[30] Kislat F and Krawczynski H 2015 *Phys. Rev. D* **92** 045016
Kostelecký V A and Mewes M 2013 *Phys. Rev. Lett.* **110** 201601
Vasileiou V *et al.* 2013 *Phys. Rev. D* **87** 122001
[31] Wei J-J *et al.* 2017 *Ap. J.* **842** 115
[32] Kostelecký V A and Mewes M 2007 *Phys. Rev. Lett.* **99** 011601
[33] LIGO Scientific Collaboration and Virgo Collaboration, Abbot B P *et al.* 2016 *Phys. Rev. Lett.* **116** 061102
[34] LIGO Scientific Collaboration and Virgo Collaboration, Abbot B P *et al.* 2016 *Phys. Rev. Lett.* **116** 241103
[35] LIGO Scientific Collaboration and Virgo Collaboration, Abbot B P *et al.* 2017 *Phys. Rev. Lett.* **118** 221101
[36] Moore G D and Nelson A E 2001 *J. High Energy Physics* JHEP09(2001)023
[37] Kostelecký V A and Mewes M 2002 *Phys. Rev. D* **66** 056005
[38] Cornish N, Blas D and Nardini G 2017 *Phys. Rev. Lett.* **119** 161102
[39] Hees A *et al.* 2016 Universe **2** 4

IOP Conf. Series: Journal of Physics: Conf. Series **952** (2018) 012005 doi:10.1088/1742-6596/952/1/012005

Structural aspects of Lorentz-violating quantum field theory

M Cambiaso[1], R Lehnert[2] and R Potting[3]

[1] Departamento de Ciencias Físicas, Universidad Andres Bello, Av. Sazié 2212, 7° piso, Santiago, Chile
[2] Indiana University Center for Spacetime Symmetries, Bloomington, Indiana 47405, USA
[3] CENTRA and Departamento de Fisica, FCT, Universidade do Algarve, 8005-139 Faro, Portugal

E-mail: mcambiaso@unab.cl

Abstract. In the last couple of decades the Standard Model Extension has emerged as a fruitful framework to analyze the empirical and theoretical extent of the validity of cornerstones of modern particle physics, namely, of Special Relativity and of the discrete symmetries C, P and T (or some combinations of these). The Standard Model Extension allows to contrast high-precision experimental tests with posited alterations representing minute Lorentz and/or CPT violations. To date no violation of these symmetry principles has been observed in experiments, mostly prompted by the Standard-Model Extension. From the latter, bounds on the extent of departures from Lorentz and CPT symmetries can be obtained with ever increasing accuracy. These analyses have been mostly focused on tree-level processes. In this presentation I would like to comment on structural aspects of perturbative Lorentz violating quantum field theory. I will show that some insight coming from radiative corrections demands a careful reassessment of perturbation theory. Specifically I will argue that both the standard renormalization procedure as well as the Lehmann-Symanzik-Zimmermann reduction formalism need to be adapted given that the asymptotic single-particle states can receive quantum corrections from Lorentz-violating operators that are not present in the original Lagrangian.

1. Introduction

The precision and sensitivity of the experiments probing potential violations of Lorentz and/or CPT symmetries have increased astonishingly [1]. This has allowed for stringent bounds on part of the parameter space of the Standard-Model Extension (SME) [2, 3] and in turn, it calls for better predictions from the model. Thus, one would like to go beyond tree-level to compute the ensuing phenomenology and contrast these predictions with experiments. Of course, this phenomenological motivation is not the only motivation for this work. We also consider it of utmost importance to consolidate the SME as a consistent theoretical framework to unravel the status of fundamental symmetry principles and therefore, without disregarding the importance of the former, we consider it crucial to contribute towards the understanding of structural aspects of quantum field theories with broken Lorentz and/or CPT symmetries. Other conceptual issues regarding perturbative quantum field theory analyses within the SME have been studied in [4, 5, 6, 7, 8, 9, 10].

In a conventional renormalizable quantum field theory, one of the implications of Lorentz symmetry is that after radiative corrections are taken into account, the Lagrangian terms that are quadratic in the fields can acquire a mass shift and field-strength factors only. Although these quantities are divergent, the latter implies that they can be dealt with systematically by the introduction of appropriate counterterms. This means that such contributions can be treated by the renormalization of quantities existing in the bare Lagrangian. In spite of the fact that even for asymptotic states self-interactions are always present, the whole construction is as if they can be disregarded altogether.

The external states spanning the asymptotic Hilbert space are crucial to properly determine the physics of free particles and equally important, to understand the meaning of the perturbative expansion upon which Feynman calculus is predicated. Perturbatively, this is seen by the fact that the propagators for the external legs, describing the quantum mechanical states of in- and out-states, are identical in structure to the momentum-space dispersion relations derived from the part of the Lagrangian that is quadratic in the fields. This provides the correct description of the propagation of free-particle states. A non-perturbative rigorous justification for this feature is given by the Lehmann–Symanzik–Zimmermann (LSZ) reduction formula [11, 12].

In the presence of Lorentz violation, specifically in the framework of the SME, previous analyses involving the properties of freely propagating particles have been performed under the tacit assumption that the latter holds. However, one may wonder whether or not this line of reasoning is still applicable. Thus the question regarding the determination of free-particle properties in the context of Lorentz-violation is structural for its description as a quantum field theory. The possibility of having finite radiative corrections and their implications in this respect will prove essential to the discussion. In this conference proceedings I wish to call the reader's attention to [13] where we shed some light on these issues.

For simplicity and on phenomenological grounds a subset of the minimal SME's electrodynamics sector will be considered. For this reason, it will suffice to work to first order in the SME coefficients. Despite the particularity of the model, it illustrates some subtleties that are worth considering.

2. The model

We start by considering the bare Lagrangian of single-flavour QED within the minimal SME.

$$\begin{aligned}\mathcal{L}_{\text{SME}} &= \tfrac{1}{2} i\,\overline{\psi}_B\,\Gamma^\mu_B \overleftrightarrow{D}^B_\mu\,\psi_B - \overline{\psi}_B\,M_B\,\psi_B - \tfrac{1}{4}(F_B)^2 \\ &\quad - \tfrac{1}{4}(k^B_F)_{\mu\nu\rho\sigma}F^{\mu\nu}_B F^{\rho\sigma}_B + (k^B_{AF})^\mu A^\nu_B \tilde{F}^B_{\mu\nu}\,. \end{aligned} \tag{1}$$

The Lorentz-violating effects are contained in the generalized gamma matrices Γ^μ_B and the generalized mass matrix M_B, which are given by:

$$\begin{aligned}\Gamma^\mu_B &= \gamma^\mu + c^{\mu\nu}_B\gamma_\nu + d^{\mu\nu}_B\gamma_5\gamma_\nu + if^\mu_B + \tfrac{1}{2}g^{\lambda\nu\mu}_B\sigma_{\lambda\nu} + e^\mu_B\,, \\ M_B &= m_B + a^\mu_B\gamma_\mu + b^\mu_B\gamma_5\gamma_\mu + \tfrac{1}{2}H^{\mu\nu}_B\sigma_{\mu\nu}\,. \end{aligned} \tag{2}$$

This flat-spacetime Lagrangian is multiplicatively renormalizable at one-loop order [6].

We will confine to the case where only $c^B_{\mu\nu}$ and $(k^B_F)_{\mu\nu\rho\sigma}$ are non-vanishing. Without loss of generality the $c^B_{\mu\nu}$ coefficient can be taken as symmetric since its antisymmetric part can be removed by a field redefinition [2, 3]. Further, we restrict the LV photon sector to that without birefringent effects. This amounts to taking $(k^B_F)^{\mu\nu\rho\sigma} = \frac{1}{2}(\eta^{\mu\rho}\tilde{k}^{\nu\sigma}_B - \eta^{\nu\rho}\tilde{k}^{\mu\sigma}_B - \eta^{\mu\sigma}\tilde{k}^{\nu\rho}_B + \eta^{\nu\sigma}\tilde{k}^{\mu\rho}_B)$, where $\tilde{k}^{\mu\nu}_B = (k^B_F)^{\mu\alpha\nu}{}_\alpha$. This choice of parameter space is made to limit the discussion to a minimum, self-consistent, and nontrivial case which nonetheless has possible phenomenological consequences. Specifically, it can be shown that, at leading order in any fermion-photon system, the $c^{\mu\nu}$ and $\tilde{k}^{\mu\nu}$ coefficients are observationally indistinguishable and only the combination $2c^{\mu\nu} - \tilde{k}^{\mu\nu}$ is measurable [14, 15, 16]. Paralleling perturbative calculations in ordinary Lorentz

IOP Conf. Series: Journal of Physics: Conf. Series **952** (2018) 012005 doi:10.1088/1742-6596/952/1/012005

symmetric case, we introduce a gauge parameter ξ and also implement a prescription for dealing with IR divergences by means of a soft-photon mass m_γ [17]. In the spirit of perturbation theory one usually chooses a zeroth-order system with known solutions, such that the remaining piece can be considered a small perturbation. Usually the zeroth-order system is chosen as the full quadratic part of the Lagrangian. In this case two reasonable choices can be made. In the first scheme, one defines as a basis the renormalized quadratic Lagrangian of the conventional Lorentz-symmetric case. All Lorentz-violating contributions to the Lagrangian are then taken as perturbations. In the second scheme, one defines as a basis the full renormalized quadratic Lagrangian, including the Lorentz-violating part, while the non-quadratic contributions to the Lagrangian are considered to be perturbations. Different advantages, drawbacks and interpretations of each scheme are described in [13]. It can be shown, however, that their predictions at one-loop level are equivalent, as it should be.

Having done this, what will be called the $c\tilde{k}$ model Lagrangian is:

$$\mathcal{L}_{c\tilde{k}} = \mathcal{L}_0 + \mathcal{L}_1 + \mathcal{L}_{c.t.}\,, \tag{3}$$

where

$$\begin{aligned} \mathcal{L}_0 &= \bar{\psi}\left[i\left(\gamma^\mu + c^{\mu\nu}\gamma_\nu\right)\partial_\mu - m\right]\psi - \frac{1}{4}\left[\eta^{\mu\nu} + \tilde{k}^{\mu\nu}\right]\left[\eta^{\alpha\beta} + \tilde{k}^{\alpha\beta}\right]F_{\mu\alpha}F_{\nu\beta} \\ &\quad -\frac{1}{2\xi}\left[\partial_\mu A^\mu + \tilde{k}^{\mu\nu}\partial_\mu A_\nu\right]^2, \end{aligned} \tag{4}$$

$$\mathcal{L}_1 = \bar{\psi}\left[-e\left(\gamma^\mu + c^{\mu\nu}\gamma_\nu\right)A_\mu\right]\psi\,. \tag{5}$$

The counterterm Lagrangian has been omitted for brevity, see [13] for details.

3. Fermion two-point function and self-energy

The quantity of concern for the determination of external fermion states is the on-shell limit of the two-point function. In this case it adopts the form: $\Gamma^{(2)}(p) = \Gamma^\mu p_\mu - m - \Sigma(p^\mu)$, where $\Gamma^\mu = \gamma^\mu + c^{\mu\nu}\gamma_\nu$. The one-loop determination of $\Sigma(p^\mu)$ will allow us to find the corresponding equation for the interacting spinor w up to that order. If in the conventional case this is $P_0 w_0 = 0$, with $P_0 = \not{p} - m$ and w_0 is the free spinor, then the desired one-loop expressions are:

$$Z_R^{-1}\,\bar{P}(p)w = 0, \qquad \bar{P}(p) \equiv P_0 + \delta P \tag{6}$$

$$(\Gamma^{(2)}(p))^{-1} = \mathcal{Z}_R\bar{P}(p)^{-1} - R. \tag{7}$$

The first term in Eq. (7) gives the one-particle pole and the second term R encodes additional off-shell effects and possibly other multiparticle physics, though it is regular around the pole. The goal is to determine $\mathcal{Z}_R, \delta P$ and R perturbatively at one-loop and to first order in LV. Extracting the the one-particle propagator pole is nontrivial as $\Gamma^{(2)}(p)$ no longer depends on $\not{p}$ only. In fact, defining $c_\gamma^p \equiv c^{\mu\nu}p_\mu\gamma_\nu$ and similarly for $\tilde{k}_\gamma^p$ we observe that LV functions c_γ^p and $\tilde{k}_\gamma^p$ do not commute with $\not{p}$. However, a generalization of the Lorentz-symmetric case is given by

$$\Gamma^{(2)}(p) = \mathcal{Z}_R^{-1}\left((c,k)_p^p\right)\bar{P}(p) + \bar{P}(p)\,\Sigma_2(\not{p}, c_\gamma^p, \tilde{k}_\gamma^p, (c,k)_p^p)\,\bar{P}(p)\,, \tag{8}$$

in the sense that it indeed has the desired property given in Eq. (7) and also is in accordance with a general form suitable for extracting the pole from the full fermion propagator as in Ref. [18].

Separating the usual Lorentz-symmetric contributions, from the LV ones with tensor structure present in the $c\tilde{k}$ model Lagrangian, and from the LV contributions with structures not present in the original Lagrangian, the fermion self-energy can be written as:

$$\Sigma(p^\mu) = \Sigma_{\rm LI}(\not{p}) + \Sigma_{\rm LV}(p^2, c_\gamma^p, \tilde{k}_\gamma^p) + \delta\Sigma(p^\mu, c^{\mu\nu}, \tilde{k}^{\mu\nu}). \tag{9}$$

The first term on the RHS of Eq. (9) corresponds to the usual Lorentz-symmetric contribution to the fermion self-energy. On the contrary, the the last two terms contain Lorentz-violating contributions to the fermion self-energy. By considering the behaviour of electromagnetic interactions under C, P, and T [13] and by defining $c_p^p \equiv c^{\mu\nu}p_\mu p_\nu$ and similarly for $\tilde{k}_p^p$, we can write to linear order in LV:

$$\Sigma_{\rm LV}(p^2, c_\gamma^p, \tilde{k}_\gamma^p) = f_2^c(p^2)\, c_\gamma^p + f_2^{\tilde{k}}(p^2)\, \tilde{k}_\gamma^p, \tag{10}$$

$$\delta\Sigma(p^\mu, c_p^p, \tilde{k}_p^p) = f_3^c(p^2)\,\frac{c_p^p}{m} + f_4^c(p^2)\,\frac{\not{p}c_p^p}{m^2} + f_3^{\tilde{k}}(p^2)\,\frac{\tilde{k}_p^p}{m} + f_4^{\tilde{k}}(p^2)\,\frac{\not{p}\tilde{k}_p^p}{m^2}, \tag{11}$$

where the functions $f_i^c(p^2)$ and $f_i^{\tilde{k}}(p^2)$ are calculable to any order in the fine-structure constant α. The difference between these two LV contributions is that the terms in $\Sigma_{\rm LV}(p^2, c_\gamma^p, \tilde{k}_\gamma^p)$ are of a form that is already contained in the original (bare) Lagrangian, while those in from $\delta\Sigma(p^\mu, c_p^p, \tilde{k}_p^p)$ are not. To proceed we will choose the second perturbation scheme described above Eq. (3) to perform a one-loop computation. The corresponding Feynman rules are:

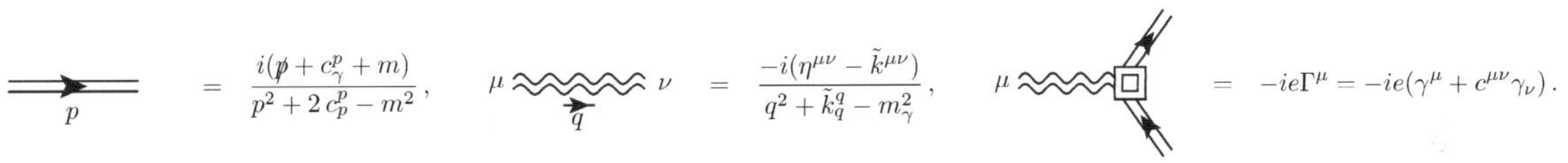

Figure 1. Feynman Rules in the $\xi = 1$ gauge.

4. Modified fermion kinetic operator and physically measurable quantities

From the one-loop expressions for the $f_i^c(p^2)$ and $f_i^{\tilde{k}}(p^2)$ functions obtained in [13] we determine the quantities we are looking for, i.e. $\mathcal{Z}_R$ and $\bar{P}$:

$$\mathcal{Z}_R^{-1} = 1 - \frac{\alpha}{\pi}\left[\ln\left(\frac{m}{m_\gamma}\right) - 1 + \frac{\gamma_E}{4} + \frac{1}{4}\ln\left(\frac{m^2}{4\pi\mu^2}\right)\right] - \frac{2\alpha}{3\pi m^2}(2\,c_p^p - \tilde{k}_p^p). \tag{12}$$

$$\bar{P}_1 = \not{p} + (c_{\rm ph})_\gamma^p - m_{\rm ph} + \frac{\alpha}{3\pi m}[2(c_{\rm ph})_p^p - (\tilde{k}_{\rm ph})_p^p]. \tag{13}$$

These are expressed in terms of the mass renormalization (which is as usual) and the corresponding expressions for the physically measurable LV coefficients, being:

$$m_{\rm ph} = m + \frac{\alpha}{\pi}\left[1 - \frac{3\gamma_E}{4} - \frac{3}{4}\ln\left(\frac{m^2}{4\pi\mu^2}\right)\right]m. \tag{14}$$

$$c_{\rm ph}^{\mu\nu} = c^{\mu\nu} - \frac{\alpha}{3\pi}\left[\frac{29}{12} - \gamma_E - \ln\left(\frac{m^2}{4\pi\mu^2}\right)\right](2c^{\mu\nu} - \tilde{k}^{\mu\nu}). \tag{15}$$

Strictly speaking we have not determined the physically measurable coefficient $\tilde{k}_{\rm ph}^{\mu\nu}$. As it differs from $\tilde{k}^{\mu\nu}$ by an order α expression we can substitute it in Eq. (13) consistently to first order in α.

The $\mathcal{O}(\alpha)$ LV radiative correction to the one-loop modified dispersion relation for asymptotic states is read off from the modified Dirac operator of Eq. (13). We observe that it is indeed proportional to $(2c_p^p - \tilde{k}_p^p)$ as required. Of course, we also find the usual shifts of conventional quantities as seen in Eqs. (12) and (14). This time, however, the fermion wave-function renormalization is no longer a constant but rather depends on the LV coefficients under consideration and on the external momentum. In addition to this we find asymptotic states in perturbative LV quantum field theory acquire higher-derivative μ-independent structures. These are the finite radiative corrections of tensor structures that were not present in the original Lagrangian referred to above and constitutes one of the key findings of this work. They stem from the $(c_{\rm ph})^p_p$ and $(\tilde{k}_{\rm ph})^p_p$ terms respectively. These features, by the way, are compatible with the general form of the one-particle fermion pole of the Källén-Lehmann representation derived in Ref. [18].

5. Asymptotic dispersion relation, the LSZ formula and S-matrix elements

The dispersion relation of asymptotic free-fermion states derived from the above is:

$$p^2 + 2(c_{\rm ph})^p_p - m^2_{\rm ph} + \frac{2\alpha}{3\pi}\left[2(c_{\rm ph})^p_p - (\tilde{k}_{\rm ph})^p_p\right] = 0. \tag{16}$$

In the Lorentz-symmetric case, scattering amplitudes are computed from on-shell external physical states. Presently, this is particularly delicate as the contributions of the finite radiative corrections whose structures are not present in the starting theory have no parallel in the usual derivation of S-matrix elements from the correlation functions. This leads to an adaptation of the LSZ reduction formula. This computation is too lengthy to reproduce here but we can outline its derivation and comment on its implications. Once in- and out- modified spinors are obtained a canonical quantization can be carried over and the Fourier expansion of the in- and out- fields are obtained from the correspondingly modified creation and annihilation operators. Next, the modified Feynman propagator can be obtained as the vacuum to vacuum time-ordered product of two in-field operators. As the wave-function renormalization is now dependent on LV quantities and also on the momentum of the external particle, this alters the usual "adiabatic" hypothesis, which relates the interacting-field operator at times much before/after the interaction takes place to the free-field operator. However, while considering all this, one can still express the fundamental scattering amplitude $\langle f|i\rangle$ in terms of vacuum to vacuum transition amplitudes of time-ordered products of interacting field operators and modified Dirac operators acting on the latter, together with wavefunction renormalization coefficients that account for self-interactions of the external states. In practical calculations, it is most useful to express the connected scattering amplitude in terms of truncated Green's functions as:

$$\begin{aligned}\langle f|i\rangle_c &= (2\pi)^4\delta^4(\sum p_i + \sum p_i' - \sum q_i - \sum q_i')\cdots \\ &\times (-i)\mathcal{Z}_R^{\frac{1}{2}}((c,\tilde{k})^{q_1}_{q_1})\bar{u}_{in}(\vec{q}_1)\cdots i\mathcal{Z}_R^{\frac{1}{2}}((c,\tilde{k})^{p_1'}_{p_1'})\bar{v}_{in}(\vec{p}_1{}') \\ &\times G^{(2n)}_{\rm trunc}(-q_1',\ldots,p_1,\ldots;-q_1,\ldots,p_1',\ldots) \\ &\times u_{in}(\vec{p}_1)(-i)\mathcal{Z}_R^{\frac{1}{2}}((c,\tilde{k})^{p_1}_{p_1})\cdots v_{in}(\vec{q}_1{}')i\mathcal{Z}_R^{\frac{1}{2}}((c,\tilde{k})^{q_1'}_{q_1'})\cdots .\end{aligned} \tag{17}$$

Eq. (17) embodies the Feynman rules for the scattering amplitude, incorporating:

- a momentum conserving delta-function;
- the amputated Green's function;
- a momentum-dependent wave-function-renormalization factor $\pm i\mathcal{Z}_R^{\frac{1}{2}}((c,\tilde{k})^p_p)$ for every external leg;
- a Dirac spinor for every external leg:
  - $u_{in}(\vec{p},\sigma)$ for an incoming fermion;
  - $\bar{u}_{in}(\vec{p},\sigma)$ for an outgoing fermion;
  - $v_{in}(\vec{p},\sigma)$ for an outgoing anti-fermion;
  - $\bar{v}_{in}(\vec{p},\sigma)$ for an incoming anti-fermion.

6. Discussion and conclusions

In this presentation we expect to have contributed to providing a detailed and thorough definition of perturbation theory in QFTs with a Lorentz-violating background. To do so we have focused on the on-shell limit of free-particle states, to first order in LV and the fine-structure constant α. A detailed computation of the fermion two-point function confirms our result is in agreement with other non-perturbative derivations [18]. Furthermore, the perturbative computations passed several cross-checks as the results were confirmed by two different perturbation schemes whose equivalence is not obvious from the beginning. Some results were also confirmed to be independent of the regularization method (by employing dimensional regularization and Pauli-Villars regularization. Also the relevant Ward-Takahashi identities were confirmed. We show a systematic reassessment of the adapted LSZ reduction formula that ultimately allows to express S-matrix elements as a series expansion in the fine-structure constant α with the corresponding set of Feynman rules to associate an amplitude to each term in the series. To illustrate our findings in [13] we analyze the non-trivial situation of infra-red divergences appearing in the scattering of a fermion off a stationary charge (Mott scattering). These issues endorse a laborious computation and allow us to summarize our conclusions.

(i) Asymptotic single-particle states receive concrete modifications due to radiative corrections. Contrary to the Lorentz symmetric case, some corrections appear as finite terms that were not in the original Lagrangian. Not all sub-sectors of the SME will necessarily suffer the same consequences, but at least it is the case for the model Lagrangian chosen, which is of considerable phenomenological relevance. However, one can expect that a more general LV theory or one with a wider parameter space might need to consider our formalism.

(ii) The wavefunction renormalization function depends on LV coefficients and on external momentum. The propagator pole is similarly altered, however, the pole structure of the fermion two-point function is expressible in a suitable manner, yielding appropriate interpretations for the mass-shell condition and renormalization conditions.

(iii) Non-trivial cancellation of IR divergences is also achieved in the fermion's dispersion relation and wavefunction renormalization function. This is illustrated for the Mott scattering cross section when the contribution of soft-photons from the external legs is considered.

(iv) All effects obtained up to the order considered are consistent with field redefinition that demands that only the combination $(2c^{\mu\nu} - \tilde{k}^{\mu\nu})$ is observable.

Acknowledgments

The author thanks support from Universidad Andrés Bello under Grant No. DI-27-11/R, DI-33-17/RG, AAEC 2017/2, FONDECYT under Grant No. 11121633 as well as the organizing committee.

References

[1] Kostelecký V A and Russell N 2011 *Rev. Mod. Phys.* **83** 11 (*Preprint* 0801.0287v10 [hep-ph])
[2] Colladay D and Kostelecký V A 1997 *Phys. Rev.* D **55** 6760 (*Preprint* hep-ph/9703464)
[3] Colladay D and Kostelecký V A 1998 *Phys. Rev.* D **58** 116002 (*Preprint* hep-ph/9809521)
[4] Kostelecký V A and Lehnert R 2001 *Phys. Rev.* D **63** 065008 (*Preprint* hep-th/0012060)
[5] Adam C and Klinkhamer F R 2003 *Nucl. Phys.* B **657** 214 (*Preprint* hep-th/0212028])
[6] Kostelecký V A, Lane C D and Pickering A G M 2002 *Phys. Rev.* D **65** 056006 (*Preprint* hep-th/0111123)
[7] Colladay D and McDonald P 2007 *Phys. Rev.* D **75** 105002 (*Preprint* hep-ph/0609084)
[8] Colladay D and McDonald P 2008 *Phys. Rev.* D **77** 085006 (*Preprint* 0712.2055 [hep-ph])
[9] Colladay D and McDonald P 2009 *Phys. Rev.* D **79** 125019 (*Preprint* 0904.1219 [hep-ph])
[10] Ferrero A and Altschul B 2011 *Phys. Rev.* D **84** 065030 (*Preprint* 1104.4778 [hep-th])
[11] Lehmann H, Symanzik K and Zimmermann W 1955 *Nuovo Cimento* **6** 319
[12] Lehmann H, Symanzik K and Zimmermann W 1957 *Nuovo Cimento* **1** 205
[13] Cambiaso M, Lehnert R and Potting R 2014 *Phys. Rev.* D **90** (*Preprint* 1401.7317 [hep-th])

IOP Conf. Series: Journal of Physics: Conf. Series **952** (2018) 012005 doi:10.1088/1742-6596/952/1/012005

[14] Colladay D and McDonald P 2002 *J. Math. Phys.* **43** 3554 (*Preprint* hep-ph/0202066)
[15] Lehnert R 2006 *Phys. Rev.* D **74** 125001 (*Preprint* hep-th/0609162)
[16] Altschul B 2006 *J. Phys. A* **39** 13757 (*Preprint* hep-th/0602235)
[17] Cambiaso M, Lehnert R and Potting R 2012 *Phys. Rev.* D **85** 085023 (*Preprint* 1201.3045 [hep-th])
[18] Potting R 2012 *Phys. Rev.* D **85**, 045033 (*Preprint* 1112.5739 [hep-th])

IOP Conf. Series: Journal of Physics: Conf. Series **952** (2018) 012006 doi:10.1088/1742-6596/952/1/012006

Matter-antimatter asymmetry in the universe via string-inspired CPT violation at early eras

Nick E Mavromatos

Theoretical Particle Physics and Cosmology Group, Physics Department, King's College London, Strand, London, WC2R 2LS, UK

E-mail: `nikolaos.mavromatos@kcl.ac.uk`

Abstract. In four-space-time dimensional string/brane theory, obtained either through compactification of the extra spatial dimensions, or by appropriate restriction to brane worlds with three large spatial dimensions, the rich physics potential associated with the presence of non-trivial Kalb-Ramond (KR) axion-like fields has not been fully exploited so far. In this talk, I discuss a scenario whereby such fields produce spontaneous Lorentz- and CPT-violating cosmological backgrounds over which strings propagate, which in the early Universe can lead to Baryogenesis through Leptogenesis in models with heavy right-handed neutrinos.

1. Introduction and Motivation

Despite its enormous success, the Standard Model (SM) of particle physics fails to provide an explanation as to why we exist, more specifically, why matter dominates over antimatter in the Universe. According to observations of the Cosmic Microwave Background (CMB) radiation [1], the baryon asymmetry is measured to be:

$$Y_{\Delta B} \equiv \frac{n_B - n_{\bar{B}}}{n_\gamma} = (6.1 \pm 0.3) \times 10^{-10} \tag{1}$$

where n_B is the number density of baryons, $n_{\bar{B}}$ is the number density of antibaryons and n_γ is the density of photons.

The baryon asymmetry (1) cannot be explained *quantitatively* within the SM. Indeed, the latter is a theory that predicts that matter and antimatter have been created in equal amounts after the Big-Bang, as a consequence of the CPT theorem [2]. According to it, the discrete symmetry transformations of Charge Conjugation (C), Parity (Spatial Reflexion, P) and Reversal in Time (antiunitary, T), when applied in any order, leave the action (and hence all dynamics derived from it) invariant, provided that the corresponding quantum field theory is Lorentz invariant, has interactions that respect locality and is unitary, that is, it has a Hermitian Hamiltonian (and the associated scattering matrix is well defined). To explain the observed matter antimatter asymmetry in the Universe in a way consistent with the above theorem, Sakharov [3] postulated that any local CPT invariant quantum field theory that was to describe the early Universe and the observed baryon asymmetry must have interactions that *violate* Baryon Number (B), Charge Conjugation Symmetry (C) and Charge-Parity (CP) symmetry. There should also be a departure from thermal equilibrium, since in a CPT invariant theory in thermal equilibrium any Baryon number violation will be washed out as a consequence of

IOP Conf. Series: Journal of Physics: Conf. Series **952** (2018) 012006 doi:10.1088/1742-6596/952/1/012006

the Boltzmann exponential factor. Although the SM contains qualitatively all such features, for instance, Baryon number is violated due to quantum anomalies, and CP Violation (CPV) appears as a complex phase in the quark mixing matrix of the weak interactions[1], nevertheless the generated matter-antimatter asymmetry in the SM is smaller than the observed one by several orders of magnitude [5, 6]. Hence, to explain the observed dominance of matter over antimatter in the Cosmos we should look for Physics beyond the SM (BSM), such as supersymmetry, extra dimensional models, string theory *etc.*, in an attempt to identify the necessary extra sources of CPV that would produce the asymmetry between baryons and antibaryons.

In fact, the observed baryon asymmetry can be generated either directly, through CPV out-of-thermal-equilibrium processes in the baryon sector in the early Universe (Baryogenesis) [7], or first through out-of-thermal-equilibrium CPV processes in the Lepton sector, often associated with CPV decays of heavy right-handed neutrinos (RHN) into SM particles, which produce a Lepton asymmetry (Leptogenesis) [8, 9]. The latter is then communicated to the baryon sector through sphaleron processes [5] that violate both Lepton (L) and Baryon (B) numbers but conserve the difference B-L. A minimal extension of the standard model, with two or three generations of heavy Majorana RHN, with a Higgs portal that allows for CPV decays of the heavy neutrinos into lepton-Higgs pairs in the early Universe, suffices to generate baryogenesis through leptogenesis [10, 11]. It is important to realise that in such scenarios the required CPV in the neutrino sector appears in the relevant mixing matrix, and as such the model cannot work with a single generation of RHN. Nevertheless, the presence of more than one generation of heavy neutrinos is required by the see-saw mechanism which is responsible for generating masses for the light neutrinos of the SM sector [12]. The difference in the widths of the decay of the heavy (Majorana) neutrinos into SM leptons as compared with the corresponding decays into antileptons, which is essential for leptogenesis, occurs at one loop level in these models. We should remark at this point, however, that there is a fair amount of fine tuning of parameters in the pertinent models that needs to be done in order for the widths of the appropriate CPV decays to produce phenomenologically realistic values for baryogenesis.

A fundamental paradigm shift for the generation of baryon asymmetry would be to relax CPT Invariance in the very early Universe. In a simplest setting, several models that include *spontaneous* Violation of Lorentz symmetry (LV) may imply CPT Violation (CPTV) [13], provided that the other two requirements of the CPT theorem, locality and unitarity[2], are obeyed. Such Lorentz- and CPT-Violating models can be parameterised in the context of the so-called Standard Model Extension (SME) [15]. Today, there are very stringent bounds on the amount of the allowed LV and CPTV, as inferred by a plethora of diverse precision measurements [16]. However, in the early Universe such effects could be much stronger, and in fact it was proposed in [17] that baryogenesis can occur within the context of string-inspired SME models [18], in which structures corresponding to tensorial quantities of order-k, which acquire non-trivial vacuum expectation values (v.e.v), $\langle T^{(k)}_{\mu_1 \dots \mu_k}\rangle \neq 0$, provide a sort of chemical potential μ and energy splitting between particles and antiparticles (quarks-antiquarks specifically). For instance, such string-inspired SME models include interactions of the form $\mathcal{L}_I \ni \frac{\lambda\,\langle T^{(k)}\rangle}{M_s^k}\overline{\psi}(\gamma^0)^{k+1}(i\partial_0)^k\psi$+h.c, where ψ are fermion fields (e.g. quarks), M_s is some high (e.g. string) mass scale and it is expected generically that $\langle T^{(k)}\rangle = \mathcal{O}\Big((m_{\rm low}/M_s)^\ell\, M_s\Big)$, where ℓ is some positive power, and $m_{\rm low}$ is a characteristic low-energy scale. Dominant terms have $k+\ell = 2$. As argued in [17], such interactions can generate a difference between the thermal equilibrium densities of quarks and antiquarks[3], e.g. for the case $k = 0, \ell = 2$: $\frac{n_q - n_{\bar q}}{s} = \frac{45\, g_q}{2\pi^4\, g_s(T)}\frac{\mu}{T}\, I_0(m_q/T)$,

[1] CP violation was observed in fact for the first time in the meson sector in neutral Kaon experiments [4].There may also be CP violation in the lepton sector (neutrino mixing matrix) but this is still unobserved.

[2] The relaxation of any one of these requirements alone would constitute independent reasons for CPTV [14].

[3] We note that if CPT is violated, the Sakharov condition of out-of-equilibrium processes can be alleviated.

IOP Conf. Series: Journal of Physics: Conf. Series **952** (2018) 012006 doi:10.1088/1742-6596/952/1/012006

with $I_0(x) = \int_x^\infty dy\, y\, \sqrt{y^2 - x^2}\, \frac{e^y}{(1+e^y)^2}$, g_q the number of internal quark degrees of freedom, $s = \frac{2\pi^2}{45}\, g_s(T)\, T^3$ the entropy density and $g_s(T) = \sum_{\rm B(osons)} g_B (\frac{T_B}{T})^3 + \frac{7}{8} \sum_{F(ermions)} g_F\, (\frac{T_F}{T})^3$ the total entropy degrees of freedom, assuming standard cosmology, with $T_{B(F)}$ the corresponding decoupling temperatures, and $g_{B(F)}$ the degrees of freedom, of Boson (Fermion) species. The induced chemical potential in this case is of order $\mu \sim m^2_{\rm low}/M_s$. Such considerations have been argued [17] to lead to a baryon asymmetry which is strong in the early Universe, but diluted today by means of sphaleron processes. The model is interesting, although, rather generic in the sense that the tensor fields which acquire non trivial v.e.v have not been specified within phenomenologically realistic microscopic string models.

2. A Lorentz and CPT Violating String-Inspired Model for Leptogenesis

It is the purpose of this talk to describe a class of detailed, and phenomenologically realistic, string-inspired quantum field theory models which can lead to efficient baryogenesis through leptogenesis within conventional string theory frameworks [19]. The latter are characterised by *spontaneous* Lorentz and CPT Violation through propagation of strings over specific LV and CPTV backgrounds that satisfy the world-sheet conformal invariance conditions and hence constitute acceptable vacua of string theory. The backgrounds are of cosmological type, that is the corresponding fields depend only on the cosmic time [20]. The background field that yields the spontaneous Lorentz and CPT Violation is the spin-one antisymmetric tensor Kalb-Ramond (KR) field $B_{\mu\nu} = -B_{\nu\mu}$ of the bosonic massless gravitational string multiplet [21, 22, 23]. Its field strength is a totally antisymmetric three-tensor $H_{\mu\nu\rho} = \partial_\mu B_{\nu\rho} +$ (cyclic permutation of the indices). In four space-time dimensions, this field strength is dual to a pseudoscalar (KR "axion") field $b^{\rm KR}(x)$: $H_{\mu\nu\rho} = e^{2\Phi} \epsilon_{\mu\nu\rho}{}^{\sigma}\, \partial_\sigma(b^{\rm KR})$, where $\epsilon_{\mu\nu\rho\sigma}$ is the covariant Levi-Civita tensor (in the presence of gravity), and Φ is the dilaton field (scalar field of the bosonic massless gravitational multiplet of the string). In the absence of other matter fields, there is a consistent cosmological solution [20] of a linearly(with the cosmic time t)-expanding Robertson-Walker Universe, which is valid to all orders in the string Regge slope $\alpha' = 1/M_s^2$. In the Einstein frame, where the (lowest-order in derivatives) scalar curvature term in the effective action is canonically normalised as in the Einstein -Hilbert action, the solution is characterised by a logarithmically (in t) dependent dilaton and a linear in t KR axion field [20]

$$b^{\rm KR}(t) = c_1\, t \tag{2}$$

where the proportionality constant c_1 depends on microscopic data of the underlying string theory. To first order in α', that is up to and including second order derivative terms of the various fields, which we restrict our attention to here, the target-space effective action obtained from this string theory is[4] [19, 20, 21, 22, 23]:

$$\mathcal{S} = \int d^4x\, \sqrt{-g}\Big[\frac{1}{2\kappa^2} R - \frac{1}{2}(\partial_\mu b^{\rm KR})^2 - \Omega + \sum_i [\overline{\psi}_i (i\gamma^\mu \partial_\mu - m_i)\psi_i + \frac{\kappa}{3\sqrt{6}} \partial_\mu b^{\rm KR}\, \overline{\psi}_i \gamma^\mu \gamma^5 \psi_i] + \dots\Big] \tag{3}$$

where the sum is over fermion species in the theory, including those in the SM sector, $\kappa = 8\pi M_P^{-2}$ is the gravitational constant in four space-time dimensions, with M_P the Planck mass scale, and Ω is a vacuum energy arising from (a plethora of terms of) the underlying microscopic string theory. In several string/brane models of the Universe, the vacuum energy Ω contains negative (anti de Sitter) parts (e.g. coming from bulk contributions in brane world scenarios [24]), which can be used to suppress the positive vacuum energy due to the kinetic terms of the KR axion

[4] From now on we also assume that the dilaton has been stabilised by its (string-loop induced) potential to an appropriate constant value, which by normalisation we set to $\Phi = 0$.

in the background (2). Such suppression is important in ensuring standard cosmology during the radiation era, a result which is used in [19] and here. The coupling of the KR axion field $b^{\rm KR}(x)$ with the fermions is dictated by the interpretation of the field strength $H_{\mu\nu\rho}$ as a totally antisymmetric part of a torsion in a generalised connection, given that the kinetic terms of $b^{\rm KR}(x)$ field and the curvature terms R in the effective action (3) can be expressed in terms of a generalised curvature scalar of such a torsionful connection [21, 22, 23]. The dots $(\ldots)$ in (3) denote gauge and other terms in the effective theory pertaining to the SM sector, but also higher derivative terms arising from strings, which are ignored for our low energy purposes.

In the presence of fermionic and other matter, the background solution (2), which implies a constant field strength, $H_{ijk} = {\rm const}, i, j, k = 1, 2, 3$, is a valid solution provided there is a constant temporal axial vector condensate of fermions that violates Lorentz and CPT symmetry but respects the isotropy of the Universe [19],

$$\sum_i \langle \overline{\psi}_i \gamma^0 \gamma^5 \psi_i \rangle \neq 0 \, . \tag{4}$$

In case of a constant $\partial_t b^{\rm KR} \equiv \dot{b}^{\rm KR}$ background, the effective action in the fermionic sector acquires the form of a SME with a non-trivial LV and CPTV axial vector background over which fermions propagate, $b_\mu \sum_i \overline{\psi}_i \gamma^\mu \gamma^5 \psi_i$, with $b_\mu \equiv \partial_\mu b^{\rm KR}$ (with only $b_0 \neq 0$ [5]). The model of [19] includes a single species of massive Majorana RHN, N_R, which also couple to the KR axion via the axial vector fermion coupling of (3). However, for such a Majorana neutrino any axial condensate vanishes, in contrast to other chiral fermions of the SM, including left handed active neutrinos, which contribute to (4). The RHN part of the effective action in the KR axion background (2) reads:

$$\frac{\mathcal{L}^N}{\sqrt{-g}} = i\overline{N_R}\gamma^\mu\partial_\mu N_R - \frac{M}{2}(\overline{N_R^c}N_R + \overline{N_R}N_R^c) + b_\mu \overline{N_R}\gamma^\mu\gamma^5 N_R - y_1\overline{L}_1\tilde{\phi}N_R + h.c.,$$

where N_R is the RHN field (with its Majorana field being given by $\nu \equiv N_R + N_R^c$), M is a Majorana mass, $\tilde{\phi}$ is the adjoint of the Higgs field, $\tilde{\phi}_i = \varepsilon_{ij}\phi_j$, and L_1 is a lepton (doublet) field of the Standard Model (SM) sector in the first generation, with y_1 a Yukawa coupling, which, if non zero, provides a non-trivial interaction between the RHN and the SM sector ("Higgs portal"). In the presence of the LV and CPTV background (2), we have (*cf.* (3)): $b_0 = \frac{\kappa}{3\sqrt{6}}\dot{b}^{\rm KR} = {\rm constant} \neq 0$. The portal yields inequivalent decay rates for the channels [19]:

$$\text{Channel I}: \quad N_R \to l^- h^+ \, , \qquad \text{Channel II}: \quad N_R \to l^+ h^- \, , \tag{5}$$

where $l^\mp$ ($h^\pm$) denote charged lepton (Higgs) fields. It is important to realise that this difference occurs already at tree level and for a single species of heavy RHN, in contrast to the conventional CPV but CPT conserving case [10, 11], where the corresponding differences due to CPV occur at one loop (*cf.* Fig. 1) and require more than one species of RHN, given that the CPV appears in the phases of the neutrino mixing matrix.

As discussed in detail in [19], on assuming $y_1 \sim 10^{-5}$ for concreteness, as well as standard cosmology during the radiation-dominance era, at which the decoupling of the heavy neutrinos occurs at a temperature $T_D \simeq M$, one can estimate the lepton asymmetry ΔL^{TOT} generated at T_D as: $\frac{\Delta L^{TOT}}{s}(T \simeq T_D \simeq M) = \frac{2\Omega b_0}{\Omega^2 + b_0^2}\frac{\bar{n}_N^{eq}(T_D)}{s(T_D)}$, where s is the entropy density, defined previously,

[5] In a quantum path integral based on the model (3), this can be seen by integrating exactly the $\partial b^{\rm KR}$ - depenendent terms, which yields [23] repulsive four-fermion interactions $J^5_\mu J^{5\,\mu}$ of the axial current $J^5_\mu = \sum_i \overline{\psi}_i \gamma_\mu \gamma^5 \psi_i$. In string effective theories, at finite temperature and density, there may be higher order (attractive) fermion contact interaction terms arising from the exchange of heavy string states, that may lead to a condensate $\langle J^5_\mu \rangle$ of the form (4) about which one expands the fermion quantum fluctuations. An SME-like effective action follows then from (3), along with (de Sitter) type effective contributions to the vacuum energy. Moreover, in this language, it is seen immediately that, upon its destruction, the condensate (being comprised of ordinary matter) will scale with the temperature as T^3 [19].

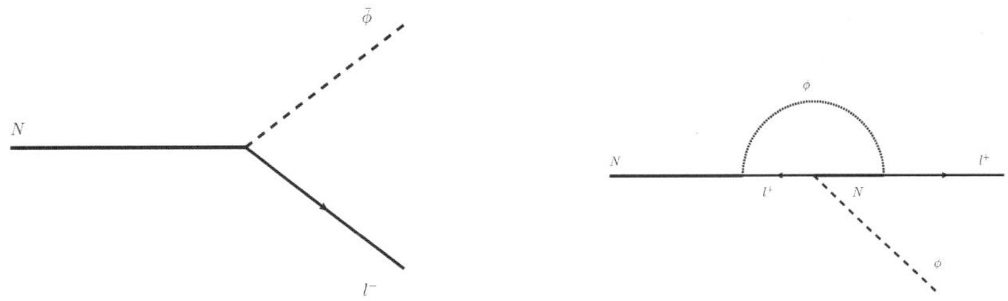

Figure 1. Tree- (left) and one-loop (right) decay amplitudes for the leptogenensis decays (5).

$\Omega^2 = M^2 + b_0^2$, and $n_N^{eq}(T) = g_N\, e^{-m_N/T}\left(\frac{MT}{2\pi}\right)^{\frac{3}{2}} + \mathcal{O}(b_0^2/M^2)$ is the thermal equilibrium density of the right-handed neutrinos, with g_N the pertinent internal degrees of freedom. We assumed that $b_0 \ll T_D, M$. This is a self consistent assumption, given that in models of baryogenesis through leptogenesis [10, 11], as in our model, the lepton asymmetry is expected to be of the same order as the baryon asymmetry (1), and in our case this is achieved for [19]:

$$\frac{b_0}{M} \simeq 10^{-8}\ , \quad \text{with} \quad M \simeq T_D \sim 10^5\ \text{GeV}, \quad y_1 \sim 10^{-5}. \tag{6}$$

In [19] a scenario was also presented for the cooling law of $\dot{b}^{\rm KR}$ with the temperature, by assuming that the axial fermion condensate (4) is destroyed at temperatures soon after the decoupling, that is $T \sim T_D$, which, on account of the equations of motion from the action (3), implies a scaling $\dot{b}^{\rm KR} \sim a(t)^{-3} \sim T^3$, for $T < T_D$, with $a(t) \sim T^{-1}$ the scale factor of the Universe during radiation dominance. Matching with the values (6), we then obtain that today the KR axion background is $b_0 = \mathcal{O}(10^{-44})$ meV, implying the the model is consistent with the current constraints of the b_μ SME parameter today [16]: $b_0 < 0.02\,\text{eV}, \quad b_i < 10^{-31}$ GeV. Moreover, as discussed in [19], the above cooling law also implies that the KR axion background is sufficiently small during the nucleosynthesis era, so the latter is not affected.

3. Discussion

In this talk we presented a string-inspired model for a LV- and CPTV-induced baryon asymmetry through leptogenesis at early eras of the Universe. Unlike the standard CPT-conserving models for leptogenesis, our model does not require mixing among RHN in order to yield the necessary amount of CPV for the generation of phenomenologically acceptable matter-antimatter asymmetry in the Universe. A single species suffices[6], provided the temporal component b_0 of the dual of the background field strength of the antisymmetric tensor field is fixed appropriately, together with the RHN mass and the Yukawa coupling of the Higgs portal (*cf.* (6)). There are many open issues of course associated with the scenario, the most important of all regarding the suppression of the vacuum energy density, which we only hinted to in this talk and in our previous work [19]. A detailed microscopic string model, where such a suppression can occur naturally, is still pending. Nonetheless, we believe that the above-presented class of string-inspired models for baryogenesis through CPTV leptogenesis constitutes an interesting avenue for future research, which has good prospects to lead to fully realistic models.

Before closing the talk I would like to make some brief remarks on the potential rôle of the quantum fluctuations of the KR axion field $b^{\rm KR}(x)$ on inducing themselves a Majorana mass term for the RHN. Such a scenario has been discussed in [25] and involves a coupling

[6] Nevertheless, we can straightforwardly extend the scenario to two or three right handed neutrinos, which then, through the seesaw mechanism [12] can give small masses to the light neutrinos of the SM sector.

(via kinetic mixing) of the KR axion with ordinary axions, $a(x)$, which in turn couple to the RHN species via chirality-preserving Yukawa coupling terms in the effective action of the form: $\mathcal{L}_a = \int d^4x\sqrt{-g}\, i\, Y_a\, a(x)\, (\overline{N}^c_R N_R - \overline{N}_R N^c_R)$. The RHN Majorana mass is then generated radiatively via two loop diagrams involving two graviton and one axion exchanges [25]. Light (SM sector) neutrino masses (compatible with the observed oscillations) can then be generated via a seesaw mechanism [12], in case there are more than one species of RHN, in a way consistent with the $\mathrm{SU_L}(2)$ gauge invariance of the SM sector. The Yukawa term $\mathcal{L}_a$ breaks the standard axion shift symmetry. In microscopic string theory models such terms may be generated by instanton effects, although the detailed microscopic dynamics is still to be explored.

Acknowledgements
This work is partially supported in part by STFC (UK) under the research grant ST/L000326/1.

References
[1] Ade, P A R, *et al.* [Planck Collaboration] 2016 Planck 2015 results. XIII. Cosmological parameters Astron. Astrophys. **594** A13.
[2] Streater R F and Wightman A S 2000 PCT, spin and statistics, and all that (Princeton, USA: Princeton Univ. Pr. (2000) 207 p.)
[3] Sakharov A D 1967 Violation of CP Invariance, c Asymmetry, and Baryon Asymmetry of the Universe Pisma Zh. Eksp. Teor. Fiz. **5** 32 [JETP Lett. **5**, 24 (1967)].
[4] Christenson J H, Cronin J W, Fitch V L and Turlay R 1964 Evidence for the 2π Decay of the K_2^0 Meson Phys. Rev. Lett. **13**, 138.
[5] Kuzmin V A, Rubakov V A and Shaposhnikov M E 1985 On the Anomalous Electroweak Baryon Number Nonconservation in the Early Universe Phys. Lett. B **155** 36.
[6] Gavela M B, Hernandez P, Orloff J, Pene O and Quimbay C. 1994 Standard model CP violation and baryon asymmetry. Part 2: Finite temperature Nucl. Phys. B **430**, 382 (1994).
[7] Buchmuller,W. 2007 Baryogenesis: 40 Years Later arXiv:0710.5857 [hep-ph].
[8] Buchmuller W, Peccei R D and Yanagida T 2005 Leptogenesis as the origin of matter Ann. Rev. Nucl. Part. Sci. **55** 311.
[9] Pilaftsis A 2009 The Little Review on Leptogenesis J. Phys. Conf. Ser. **171** 012017.
[10] M. Fukugita M and Yanagida T 1986 Baryogenesis Without Grand Unification, Phys. Lett. B **174** 45.
[11] Luty M A 1992 Baryogenesis via leptogenesis Phys. Rev. D **45** 455.
[12] G. Altarelli and K. Winter (eds.) 2003 Neutrino mass Springer Tracts Mod. Phys. **190**.
[13] Greenberg O W 2002 CPT violation implies violation of Lorentz invariance Phys. Rev. Lett. **89** 231602.
[14] Chaichian M, Dolgov A D, Novikov V A and Tureanu A 2011 CPT Violation Does Not Lead to Violation of Lorentz Invariance and Vice Versa, Phys. Lett. B **699** 177.
[15] Colladay D and Kostelecky V A 1998 Lorentz violating extension of the standard model Phys. Rev. D **58** 116002.
[16] Kostelecky V A and Russell N 2011 Data Tables for Lorentz and CPT Violation Rev. Mod. Phys. **83** 11.
[17] Bertolami O, Colladay D, Kostelecky V A and Potting R 1997 CPT violation and baryogenesis Phys. Lett. B **395** 178.
[18] Kostelecky V A and Potting R 1991 CPT and strings Nucl. Phys. B **359** 545.
[19] de Cesare M, Mavromatos N E and Sarkar S 2015 On the possibility of tree-level leptogenesis from Kalb-Ramond torsion background Eur. Phys. J. C **75** 514.
[20] Antoniadis I, Bachas C, Ellis J R and Nanopoulos D V 1989 An Expanding Universe in String Theory Nucl. Phys. B **328** 117.
[21] Gross D J and Sloan J H 1987 The Quartic Effective Action for the Heterotic String Nucl. Phys. B **291** 41.
[22] Metsaev R R and Tseytlin A A 1987 Order alpha-prime (Two Loop) Equivalence of the String Equations of Motion and the Sigma Model Weyl Invariance Conditions: Dependence on the Dilaton and the Antisymmetric Tensor Nucl. Phys. B **293**, 385 (1987).
[23] Duncan M J, Kaloper N and Olive K A 1992 Axion hair and dynamical torsion from anomalies Nucl. Phys. B **387** 215.
[24] Mavromatos N E and Rizos J 2003 Exact solutions and the cosmological constant problem in dilatonic domain wall higher curvature string gravity Int. J. Mod. Phys. A **18**, 57.
[25] Mavromatos N E and Pilaftsis A 2012 Anomalous Majorana Neutrino Masses from Torsionful Quantum Gravity Phys. Rev. D **86** 124038.

IOP Conf. Series: Journal of Physics: Conf. Series **952** (2018) 012007 doi:10.1088/1742-6596/952/1/012007

Inflation with a massive vector field nonminimally coupled to gravity

J Páramos

Centro de Física do Porto, Faculdade de Ciências da Universidade do Porto, Rua do Campo Alegre 687, 4169-007 Porto, Portugal

E-mail: jorge.paramos@fc.up.pt

Abstract. The possibility that inflation is driven by a massive vector field with $SO(3)$ global symmetry nonminimally coupled to gravity is presented. Through an appropriate *Ansatz* for the vector field, the behaviour of the equations of motion is studied through the ensuing dynamical system, focusing on the characterisation of the ensuing fixed points.

1. Introduction

Inflation is a key ingredient for solving the problems of the Cosmological Standard Model [1, 2, 3, 4, 5]. Most proposals imply that it is driven by a scalar field with a suitable potential; however, the possibility that inflation can result from the dynamics of vector fields is also interesting: this has been approached *e.g.* in Refs. [6, 7], resorting to a quadratic potential $V(A^\mu) \sim A_\mu A^\mu$, and more recently in Ref. [8] (see also Refs. [9, 10, 11, 12]).

This work describes the dynamics of a vector field as first posited in Ref. [7] (see also Refs. [10, 12, 13, 14], but with a nonminimal coupling between the vector field and both the Ricci scalar and Ricci tensor [15]. The latter is also found in the similar context posited by the so-called bumblebee vector models for spontaneous breaking of Lorentz symmetry [16, 17, 18, 19, 20].

2. The Model

We consider the action for an $SO(3)$-invariant gauge group with a massive vector field nonminimally coupled to the curvature:

$$S = \int d^4x\sqrt{-g}\left(\frac{R}{2k^2} + \frac{1}{8e^2}\mathrm{Tr}[F_{\mu\nu}F^{\mu\nu}] + \frac{1}{2}m^2\mathrm{Tr}[A_\mu A^\mu] + \frac{1}{3}\alpha R\mathrm{Tr}[A_\mu A^\mu] + \beta\mathrm{Tr}[R_{\mu\nu}A^\mu A^\nu]\right) , \tag{1}$$

where $k^2 = 8\pi G$, e is the gauge coupling, α and β are the strengths of the nonminimal couplings between the gauge field and the Ricci scalar and Ricci tensor, respectively [7]. The gauge field strength is given by $F_{\mu\nu} = \partial_\mu A_\nu - \partial_\nu A_\mu + [A_\mu, A_\nu]$.

The variation of the action with respect to the metric yields

$$\begin{aligned}\frac{1}{2k^2}\left(R_{\mu\nu} - \frac{1}{2}g_{\mu\nu}R\right) &= -\frac{1}{k^2}R_{\mu\nu} + g_{\mu\nu}\mathcal{L} - m^2A_\mu A_\nu - \frac{1}{2e^2}F_{\mu\nu}F^{\mu\nu} \\ &-\frac{2}{3}\alpha\left[R_{\mu\nu}(A^\rho A_\rho) + RA_\mu A_\nu - \nabla_\mu\nabla_\nu(A_\rho A^\rho) + g_{\mu\nu}R(A_\rho A^\rho)\right] \\ &+\beta\left[2\nabla_\beta\nabla_{(\mu}A_{\nu)}A^\beta - g_{\mu\nu}(\nabla_\alpha\nabla_\beta A^\alpha A^\beta) - \Box(A_\mu A_\nu) - 4A^\alpha R_{\alpha(\mu}A_{\nu)}\right] ,\end{aligned} \tag{2}$$

IOP Conf. Series: Journal of Physics: Conf. Series **952** (2018) 012007 doi:10.1088/1742-6596/952/1/012007

while the variation with respect to the gauge field gives the vector field equations of motion,

$$\frac{1}{8e^2}\nabla_\mu(\nabla_\nu A^\nu) + \left(\frac{1}{2}m^2 + \frac{1}{3}\alpha R\right) A_\mu + \beta R_{\mu\nu} A^\nu = 0 \ . \tag{3}$$

3. Cosmological Dynamics

We adopt the Robertson-Walker metric

$$ds^2 = -N(t)^2 (dx^0)^2 + a(t)^2 \sum_{i=1}^{3} (dx^i)^2 \ , \tag{4}$$

where $N(t)$ is the lapse function and $a(t)$ is the scale factor. From the thorough discussion of Refs. [7, 15], it is found that the vector field takes the form

$$A_0 = 0 \quad , \quad A_i = A_i^a L_a = \chi_0(t) \delta_i^a L_a \ , \tag{5}$$

with $\chi_0(t)$ being an arbitrary function and L_a the generators of the internal $SO(3)$ group.

The cosmological field equations follow from the substitution of Eqs. (4) and (5) into Eqs. (2) and (3) or, since all constraints are respected, they can also be obtained by replacing the *Ansatz* Eqs. (4) and (5) into action (1) and discarding the infinite volume of the spatial hypersurface,

$$S_{\text{eff}} = 3 \int dt \left[-\frac{a\dot{a}^2}{k^2 N} + \frac{a}{4Ne^2} \left(\frac{\dot{\chi}_0^2}{2} - \frac{N^2}{a^2} V(\chi_0) \right) + \left(\frac{1}{4} N m^2 + \gamma \frac{\dot{a}^2}{Na} \right) \chi_0^2 \right] \ , \tag{6}$$

where we define the quartic potential $V(\chi_0) = \chi_0^4/8$ and the composite coupling $\gamma \equiv \alpha - \beta$, showing that the contribution of the two couplings between the vector field and the curvature have similar dynamical impact; notice that setting $\beta = \alpha$ implies a vanishing effect from to the aforementioned couplings. In the remainder of this study, we thus consider that $\gamma \neq 0$.

Varying the above with respect to $a(t)$, $N(t)$ and $\chi_0(t)$ and setting the gauge $N(t) = 1$ yields the Friedmann and Raychaudhuri equations and the equation of motion for the vector field:

$$4(a^2 - k^2 \gamma \chi_0^2) H^2 = \frac{k^2}{e^2} \left(\frac{\dot{\chi}_0^2}{2} + \frac{V}{a^2} \right) + k^2 m^2 \chi_0^2 \ , \tag{7}$$

$$(a^2 - k^2 \gamma \chi_0^2)(\dot{H} + H^2) = -H^2 a^2 + k^2 \left(2\gamma \dot{\chi}_0 \chi_0 H + \frac{m^2 \chi_0^2}{4} \right) \ , \tag{8}$$

$$\ddot{\chi}_0 + H \dot{\chi}_0 = -\frac{\chi_0^3}{2a^2} + 8e^2 H^2 \gamma \chi_0 - 2e^2 m^2 \chi_0 \ , \tag{9}$$

where $H = \dot{a}(t)/a(t)$ is the Hubble parameter.

To find general inflationary solutions, we must solve the dynamical system associated with Eqs. (7)–(9) and physically interpret the ensuing critical points. To do so, the following dimensionless variables are introduced,

$$x = \frac{k\chi_0(t)}{a(t)\sqrt{1-w^2}} \quad , \quad y = \frac{k^2 \dot{\chi}_0(t)}{2\sqrt{2(1-w^2)}} \quad , \quad z = kH \quad , \quad \tau = \frac{t}{k} \ , \tag{10}$$

with the auxiliary function

$$w = \sqrt{\gamma} \frac{k\chi_0(t)}{a(t)} = x \sqrt{\frac{\gamma}{1+\gamma x^2}} \ . \tag{11}$$

The Friedmann Eq. (7) yields the algebraic constraint

$$z^2 = y^2 + \frac{1}{32}\frac{x^4}{1+\gamma x^2} + \frac{1}{4}\mu^2 x^2 \ , \tag{12}$$

where $\mu = mk$ is the reduced mass, and only two degrees of freedom remain. We calculate the derivative of the variables (x, y) with respect to the dimensionless time τ, obtaining

$$\begin{aligned} \frac{dx}{d\tau} &= (1+\gamma x^2)\left[y - x\sqrt{y^2 + \frac{1-w^2}{32}x^4 + \frac{1}{4}\mu^2 x^2}\right] \ , \\ \frac{dy}{d\tau} &= -\frac{1-w^2}{4\sqrt{2}}x^3 - \frac{\mu^2}{\sqrt{2}}x + \frac{4\gamma}{\sqrt{2}}\left(2y^2 + \frac{1-w^2}{32}x^4 + \frac{1}{4}\mu^2 x^2\right)x - \\ & \quad (\gamma x^2 + 2)y\sqrt{y^2 + \frac{1-w^2}{32}x^4 + \frac{1}{4}\mu^2 x^2} \ . \end{aligned} \tag{13}$$

3.1. Finite critical points

Considering the dynamical system Eq. (13), we first analyse the trivial critical point $F(0,0)$: the eigenvalues of the Jacobian matrix derived from Eq. (13) at this point are $\lambda_\pm = \pm 2i\sqrt{2}\mu$, indicating that this is a stable critical point, as expected from the vanishing of the vector field.

Eight non-trivial critical points also arise, corresponding to the possible combinations $(\pm\sqrt{X_\pm}, \pm\sqrt{Y_\pm})$, defining for convenience

$$\begin{aligned} X_\pm &= \frac{2+\mu^2 \pm \sqrt{(1-16\gamma)^2\mu^4 + 4(8\gamma+1)\mu^2 + 4}}{2\gamma\left[1+(8\gamma-1)\mu^2\right]} \ , \\ Y_\pm &= \frac{1}{12\gamma(4\gamma-1)}\left[\mu^2 + \frac{2+(1+4\gamma)\mu^2}{8}X_\pm\right] \ . \end{aligned} \tag{14}$$

Since the dynamical system is invariant under reflections $(x, y) \to (-x, -y)$, it suffices to analyse those presented in Table 1.

Table 1. Non-trivial, finite critical points.

Point	(x, y)	H	Point	(x, y, z)	H
A	$(\sqrt{X_+}, \sqrt{Y_+})$	H_+	B	$(\sqrt{X_+}, -\sqrt{Y_+})$	H_+
C	$(\sqrt{X_-}, \sqrt{Y_-})$	H_+	D	$(\sqrt{X_-}, -\sqrt{Y_+})$	H_-

Using the algebraic constraint Eq. (12) to read the Hubble parameter yields Eq. (15)

$$H_{0\pm}^2 = \frac{2+(1+8\gamma)(mk)^2}{24k^2\gamma(4\gamma-1)}\left(1 \pm \sqrt{1 + \frac{48(4\gamma-1)\gamma(mk)^4}{[2+(1+8\gamma)(mk)^2]^2}}\right) \ , \tag{15}$$

so that the vector field can be read from the definition (11),

$$A^2 = 4\left[(4\gamma-1)H_0^2 - m^2\right] = \frac{8}{k^2\left[12\gamma + \left(\frac{m}{H_0}\right)^2\right]} \ . \tag{16}$$

Following a numerical procedure to evaluate the eigenvalues of the Jacobian matrix evaluated at the critical points (A, B, C, D), one finds that these are saddle points [15]. We also find that the critical points (C, D, G, H) are unphysical, as they lead to an imaginary Hubble parameter.

IOP Conf. Series: Journal of Physics: Conf. Series **952** (2018) 012007 doi:10.1088/1742-6596/952/1/012007

It does not suffice to require that (x, y) are real-valued for the critical points (A, B) to be physically meaningful: indeed, one must also consider the definition of the dimensionless variables Eq. (10) to ensure that the physical vector field and its time derivative are well defined. This requires that $1 - w^2 > 0 \to 1 + \gamma x^2 > 0$, which together with the requirement that (x, y) are real translates into the condition $\gamma > 1/4$.

3.2. Critical points at infinity

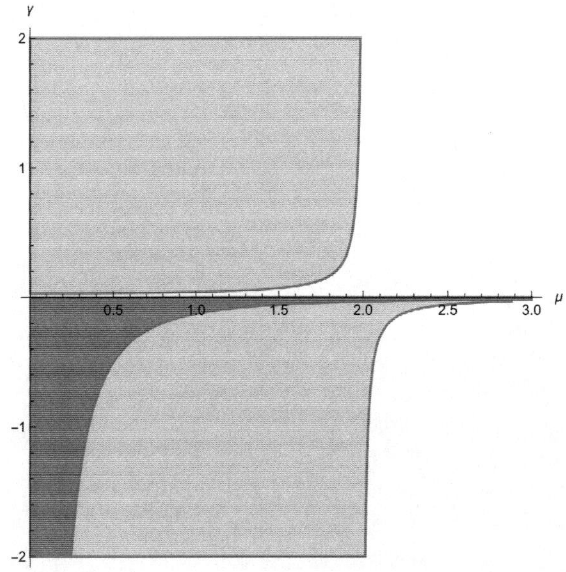

Figure 1. Degenerate eigenvalues of $S_\pm$: real and negative (dark gray), imaginary (light gray).

We now analyse putative critical points found at infinity: in order to do so, we introduce a new radial coordinate and time variable, together with the usual definition of polar angle,

$$x = \frac{\rho \cos\theta}{1-\rho} \quad , \quad y = \frac{\rho \sin\theta}{1-\rho} \quad , \quad \frac{d\zeta}{d\tau} = \frac{1}{(1-\rho)^2}, \tag{17}$$

where $0 \leq \rho \leq 1$. By considering $\rho = 1$ and approaching the ensuing dynamical system, it is possible to show that three additional fixed points arise [15], depicted in Table 2 together with the eigenvalues of the related Jacobian matrix.

Table 2. Eigenvalues of the critical points at infinity N and $S_\pm$.

Point	θ	Eigenvalues
N	$\frac{\pi}{2}$	$\frac{3\pm\sqrt{1-64\gamma}}{2}$
$S_\pm$	$\arccos\left(\pm\sqrt{\frac{1}{1-\frac{1}{32\gamma}-\frac{1}{4}\mu^2}}\right)$	$\pm\frac{8\gamma\sqrt{-\gamma(8\gamma\mu^2+1)}}{1+8\gamma(\mu^2-4)}$

3.2.1. Critical Point N Inspection of Table 2 shows that the critical point $N(1, \pi/2)$ is a saddle point, if $\gamma \leq -1/8$; unstable, if $-1/8 < \gamma < 1/64$; or a focus, if $\gamma > 64$. Furthermore, the constraint (12) shows that $H \to \infty$ in all cases, thus yielding the possibility of a Big Rip.

3.2.2. Critical Points $S_\pm$ By inspecting Table 2, we may ascertain the behaviour of the critical points $S_\pm$: we impose Eq. (12) for real critical points and vary the coupling γ and reduced mass μ to determine the behaviour of the corresponding degenerate eigenvalues, as depicted in Fig. 1. We find that the latter are never positive, yielding the constraints

$$\gamma < 0 \quad : \quad -\frac{1}{8\mu^2} < \gamma < 0 \quad , \quad \mathrm{Re}(\gamma) = 0 \quad : \quad \begin{cases} \gamma < -\frac{1}{8\mu^2} \vee \gamma > \frac{1}{8(4-\mu^2)} \ , & \mu < 2 \\ \frac{1}{8(4-\mu^2)} < \gamma < -\frac{1}{8\mu^2} \ , & \mu > 2 \end{cases} . \quad (18)$$

Again resorting to Eq. (12), we find that $H^2 = -1/[32(k\gamma)^2] < 0$, so that the fixed point N leads to an unphysical, imaginary expansion rate.

4. Conclusions

In this work, the dynamics of an $SO(3)$-invariant massive vector field [7] nonminimally coupled to the curvature are presented. The resulting system admits De Sitter solutions for a restricted region of the parameter space, $\gamma > 1/4$. The dynamical system arising from the equations of motion was studied, finding 9 finite critical points and 3 critical points at infinity. Aside from the trivial fixed point corresponding to no vector field, the other 8 non-trivial points are saddles leading to a constant Hubble parameter rate, with only 2 physical. Depending on the value of γ, it was found that the fixed point at infinity N can behave as a saddle point, an unstable point or a focus — in which case it leads to a Big Rip. The other 2 critical points at infinity $S_\pm$ yield an imaginary Hubble parameter, and can thus be identified with an oscillating universe.

Acknowledgments

The author wishes to thank the organization of *CPT and Lorentz Symmetry in Field Theory* (Faro, July 2017) for its hospitality, and O. Bertolami for fruitful discussions.

References

[1] Starobinsky A A 1980 *Phys. Lett.* B **91** 99
[2] Guth A 1981 *Phys. Rev.* D **23** 347
[3] Albrecht A and Steinhardt P J *Phys. Rev. Lett.* **48** 1220
[4] Linde A 1982 *Phys. Lett.* B **108** 389
[5] Olive K A 1990 *Physics Reports* **190** 307
[6] Ford L H 1989 *Phys. Rev.* **40** 967
[7] Bento M C, Bertolami O, Moniz P V, Mourão J M and Sá P M 1993 *Class. Quantum Gravity* **10** 2
[8] Koivisto T S and Mota D F 2008 *JCAP* **08** 021
[9] Tartaglia A and Radicella N 2007 *Phys. Rev.* D **76** 083501
[10] Golovnev A, Mukhanov V and Vanchurin V 2008 *JCAP* **0806** 009
[11] Golovnev A and Vanchurin V 2009 *Phys. Rev.* D **79** 103524
[12] Esposito-Farese G, Pitrou C and Uzan J P 2010 *Phys. Rev.* D **81** 063519
[13] Bertolami O and Mota D F 1999 *Phys. Lett.* B **455** 96
[14] Turner M S and Widrow L M 1988 *Phys. Rev.* D **37** 2743
[15] Bertolami O, Bessa V and Páramos J 2016 *Phys. Rev.* D **93** 064002
[16] Kostelecký V A 2004 *Phys. Rev.* D **69** 105009
[17] Bluhm R and Kostelecký V A 2005 *Phys. Rev.* D **71** 065008
[18] Jacobson T 2007 *PoS* **QG-PH** 020
[19] Bertolami O and Páramos J 2005 *Phys. Rev.* D **72** 044001
[20] Capelo D and Páramos J 2015 *Phys. Rev.* D **91** 104007

IOP Conf. Series: Journal of Physics: Conf. Series **952** (2018) 012008 doi:10.1088/1742-6596/952/1/012008

Constraining Lorentz Violation in Electroweak Physics

Ralf Lehnert

Indiana University Center for Spacetime Symmetries, Bloomington, IN 47405, USA
Leibniz Universität Hannover, Welfengarten 1, 30167 Hannover, Germany

E-mail: `ralehner@indiana.edu`

Abstract. For practical reasons, the majority of past Lorentz tests has involved stable or quasistable particles, such as photons, neutrinos, electrons, protons, and neutrons. Similar efforts in the electroweak sector have only recently taken shape. Within this context, Lorentz-violation searches in the Standard-Model Extension's Z-Boson sector will be discussed. It is argued that existing precision data on polarized electron–electron scattering can be employed to extract the first conservative two-sided limits on Lorentz breakdown in this sector at the level of 10^{-7}.

1. Introduction

Investigations of Lorentz symmetry may provide experimental access to physics beyond the Standard Model [1, 2, 3, 4, 5]. The effective-field-theory description of such effects is provided by the Standard-Model Extension (SME) framework [6, 7], which contains the usual Standard Model and General Relativity as special cases, but also incorporates Lorentz- and CPT-violating operators of arbitrary mass dimension. To date, the SME framework has not only been employed for numerous experimental [8, 9, 10, 11, 12, 13, 14] and theoretical [15, 16, 17, 18, 19, 20] analyses of Lorentz breakdown, but also for phenomenological studies of spacetime torsion [21] and nonmetricity [22].

Previous Lorentz-violation searches have primarily focused on stable or quasistable particles. Only a few studies have been performed in the context of weak-interaction physics [23, 24, 25]; only some of these studies have constrained the SME coefficients for the massive gauge bosons [24, 25]. The present work reports on recent progress in this field. In particular, we discuss the idea that polarized electron–electron scattering is affected by the Lorentz-breaking SME coefficients $k_{\phi\phi}$ and k_W associated with the Z boson. Both of these coefficients are CPT even and part of the minimal SME (mSME), which restricts attention to power-counting renormalizable Lorentz-breaking operators.

The presentation of this report is divided into two parts. The first part reviews the theoretical aspects of tree-level polarized Møller scattering in the mSME and derives a general expression for the corresponding cross section. The second part focuses on a special case of this cross section, in which the structure of the Lorentz violation takes a simplified form. The goal of this second part is to identify generic experimental signals and estimate the sensitivities that could be reached. Throughout, natural units $\hbar = c = 1$ are used, and the convention for the Minkowski metric are $\eta^{\mu\nu} = \mathrm{diag}\,(+,-,-,-)$.

2. Theory

The mSME is set up to describe general power-counting renormalizable Lorentz breakdown while retaining other fundamental physics properties including the ordinary SU(3)×SU(2)×U(1) gauge structure. The mSME's electroweak parameter space is therefore most transparently introduced before SU(2)×U(1) symmetry violation. Spontaneous gauge-symmetry breaking then mixes the mSME coefficients k_B, k_W, and $k_{\phi\phi}$ of the original U(1), SU(2), and Higgs sectors [6]. This leads to the following quadratic contributions to the mSME sector containing the photon and the Z boson [6, 26]:

$$\begin{aligned}\delta\mathcal{L}^{(2)}_{A,Z} = & -\tfrac{1}{4}(k_B\cos^2\theta_W + k_W\sin^2\theta_W)_{\kappa\lambda\mu\nu}\,F^{\kappa\lambda}F^{\mu\nu}\\ & -\tfrac{1}{4}(k_W\cos^2\theta_W + k_B\sin^2\theta_W)_{\kappa\lambda\mu\nu}\,Z^{\kappa\lambda}Z^{\mu\nu}\\ & -\tfrac{1}{4}\sin 2\theta_W\,(k_W - k_B)_{\kappa\lambda\mu\nu}\,F^{\kappa\lambda}Z^{\mu\nu}\\ & +\tfrac{1}{2}M_Z^2\,\mathrm{Re}\,(k_{\phi\phi})_{\mu\nu}\,Z^\mu Z^\nu\,,\end{aligned} \tag{1}$$

where $F^{\mu\nu} = \partial^\mu A^\nu - \partial^\nu A^\mu$ is now the photon gauge field corresponding to the remaining unbroken $\mathrm{U}(1)_\gamma$ symmetry, and is $Z^{\mu\nu} = \partial^\mu Z^\nu - \partial^\nu Z^\mu$ the massive gauge field corresponding to the Z boson. The usual weak angle is denoted by θ_W and the mass of the Z by M_Z.

Disregarding Chern–Simons-type Lorentz breakdown, the Lagrangian corrections (1) are general in a purely theoretical sense. However, previous observations may be employed to eliminate those coefficients whose experimental limits are well beyond the sensitivity reach of present-day and near-future Møller-scattering measurements. This arguments applies in particular to the mSME coefficient

$$(k_F)_{\kappa\lambda\mu\nu} \equiv (k_B\cos^2\theta_W + k_W\sin^2\theta_W)_{\kappa\lambda\mu\nu} \tag{2}$$

appearing in Eq. (1), which governs the photon: existing bounds effectively render k_F zero for our present purposes. Then, k_B may be expressed in terms of k_W as

$$(k_B)_{\kappa\lambda\mu\nu} = -\tan^2\theta_W\,(k_W)_{\kappa\lambda\mu\nu}\,. \tag{3}$$

This phenomenological simplification together with Eq. (1) determines the Feynman rules given in Fig. 1. These rules govern the dominant Lorentz-violating effects in electron–electron scattering.

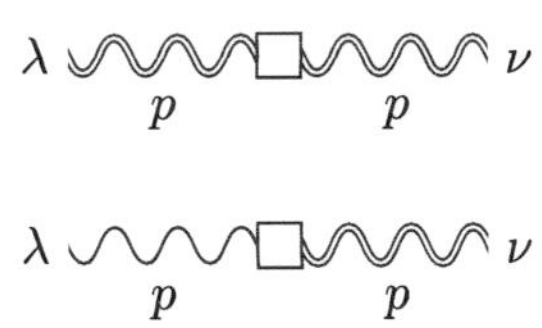

$$= -2i(1-\tan^2\theta_W)\,(k_W)_{\kappa\lambda\mu\nu}\,p^\kappa p^\mu + iM_Z^2\,\mathrm{Re}(k_{\phi\phi})_{\lambda\nu}$$

$$= -2i\tan\theta_W\,(k_W)_{\kappa\lambda\mu\nu}\,p^\kappa p^\mu$$

Figure 1. Feynman rules for Lorentz-breaking corrections to the Z boson. The corrections relevant for the present report take the form of propagator insertions. The single and double wavy lines represent the conventional Lorentz-symmetric photon and Z-boson propagators, respectively. Taken from Ref. [26].

With the Feynman rules in place, the Lorentz-violating corrections to polarized Møller scattering can be determined. This process is dominated by the electromagnetic interaction via photon exchange. However, Lorentz-invariant subdominant contributions arise through diagrams in which the internal photon line is replaced by an internal Z-boson line. Although

IOP Conf. Series: Journal of Physics: Conf. Series **952** (2018) 012008 doi:10.1088/1742-6596/952/1/012008

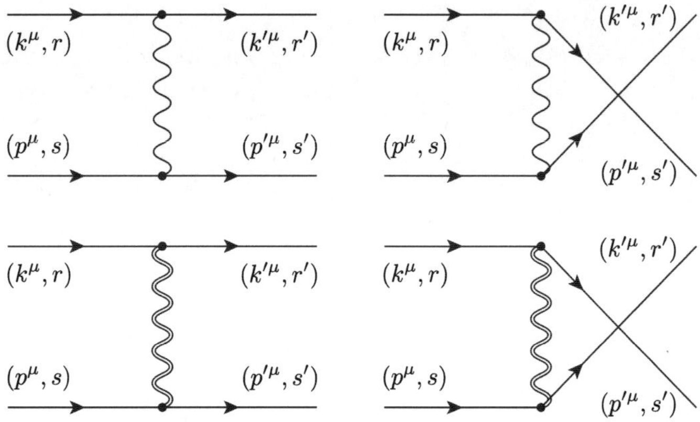

Figure 2. Leading conventional tree-level contributions to Møller scattering. Solid lines denote electrons. They are labeled by one of the four external 4-momenta $k^\mu, k'^\mu, p^\mu, p'^\mu$ as well as by one of the helicity observables r, r', s, s'. A single wavy line represents the ordinary Lorentz-symmetric photon propagator, and a double wavy line the usual Z-boson propagator. All vertices are the conventional ones. Taken from Ref. [26].

small, the associated effects have previously been measured [27] for the purpose of investigating the usual weak charge of the electron

$$Q_W^e = 4\sin^2\theta_W - 1\,, \tag{4}$$

and future, more sensitive measurements are planned [28]. But the precision of such measurements and their dependence on Z-boson physics make them also an excellent candidate for the study of the Lorentz-violating effects represented in Fig. 1.

To calculate these Lorentz-breaking Z-boson effects in polarized Møller scattering within the mSME, we make the reasonable assumption that mSME coefficients are small, and leading-order result are therefore sufficient. We have previously argued that the photon's mSME coefficients are known to be too small to contribute to observable effects in the present context and can hence be disregarded here. An analogous reasoning holds for the electron's mSME coefficients, so that we may take the leading Lorentz-breaking effects in our Møller process to be entirely due to the propagator insertions shown in Fig 1. At tree level, this process is thus described by the conventional diagrams depicted in Fig. 2 and the Lorentz-violating corrections displayed in Fig. 3.

To streamline the calculation of the diagrams, it is helpful to tailor the external-leg polarizations to the actual experimental set-up. The measurements we have in mind, such as the E158 experiment at the Stanford Linear Accelerator Center (SLAC) [27], typically focus a beam of incoming, longitudinally polarized, relativistic electrons on a fixed unpolarized target. After the scattering process, outgoing Møller electrons in a finite range of scattering angles are detected and counted. It follows that we have to perform a calculation for incoming states of definite helicity, average over the spin states of the fixed-target electrons, and sum over outgoing electron spins. This gives two squared matrix elements $\overline{|\mathcal{M}_R|^2}$ and $\overline{|\mathcal{M}_L|^2}$ for incident right-handed and left-handed beam electrons, respectively. The general structure of these is determined by

$$\begin{aligned} \overline{|\mathcal{M}_R|^2} &= \overline{|\mathcal{M}_R^0|^2} + \delta\overline{|\mathcal{M}_R|^2}\,, \\ \overline{|\mathcal{M}_L|^2} &= \overline{|\mathcal{M}_L^0|^2} + \delta\overline{|\mathcal{M}_L|^2}\,, \end{aligned} \tag{5}$$

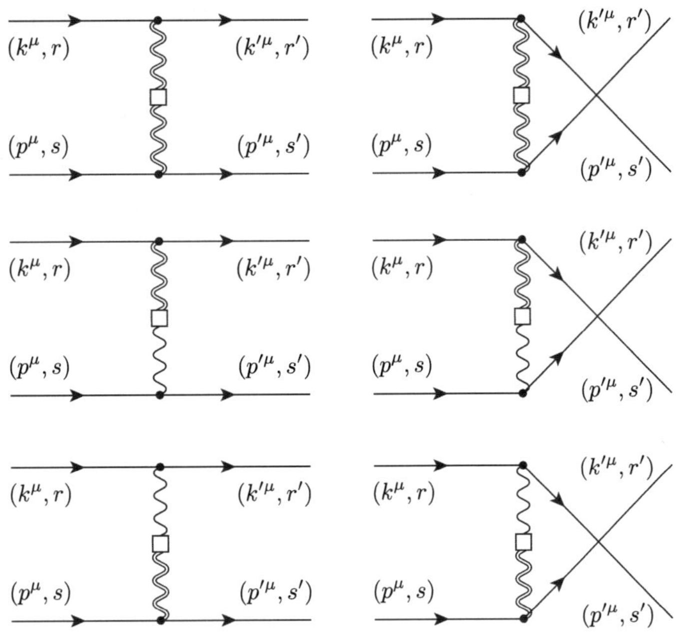

Figure 3. Dominant Lorentz-breaking corrections to Møller scattering. The notation is the same as before. In particular, the 3-point vertices are the conventional Lorentz-invariant ones, and the insertions are those shown in Fig. 1. Taken from Ref. [26].

where $\overline{|\mathcal{M}_R^0|^2}$ and $\overline{|\mathcal{M}_L^0|^2}$ are the ordinary Lorentz-symmetric contributions, which can be found in the literature [29]. The Lorentz-breaking effects in this Møller process are contained in the corrections $\delta\overline{|\mathcal{M}_R|^2}$ and $\delta\overline{|\mathcal{M}_L|^2}$. Our present interest is directed at the calculation of explicit expressions for these corrections as functions of the relevant kinematical quantities and the k_W and $k_{\phi\phi}$ coefficients.

To cast the result for $\delta\overline{|\mathcal{M}_R|^2}$ and $\delta\overline{|\mathcal{M}_L|^2}$ into a relatively compact form, we note that the four conventional external momenta k^μ, k'^μ, p^μ, and p'^μ obey energy–momentum conservation, as usual. This leaves three independent momenta, which we choose to parametrize as

$$S_\mu = k_\mu + p_\mu \,, \tag{6}$$
$$T_\mu = k'_\mu - k_\mu \,, \tag{7}$$
$$U_\mu = p_\mu - k'_\mu \,, \tag{8}$$

with a notation inspired by the Mandelstam variables. With these definition, one can show

IOP Conf. Series: Journal of Physics: Conf. Series **952** (2018) 012008 doi:10.1088/1742-6596/952/1/012008

that [26]

$$\delta\overline{|\mathcal{M}_R|^2} = \frac{2e^4(k_W)^{\kappa\mu\lambda\nu}}{s\,M_Z^2\,y^2(1-y)^2\cos^2\theta_W}\Big\{(y-1)\left[(2-4y+y^2)Q_W^e+1-2y\right]\,S_\kappa T_\mu S_\lambda T_\nu$$
$$+\,y\left[(1-2y-y^2)Q_W^e+1-2y\right]\,S_\kappa U_\mu S_\lambda U_\nu+\left[(2-y+y^2)Q_W^e+1\right]\,T_\kappa U_\mu T_\lambda U_\nu$$
$$-\,s\,y\,(1-y)\,[y\,(2-y)Q_W^e+1]\,\eta_{\mu\nu}\,T_\kappa T_\lambda-s\,y\,(1-y)\,\left[(1-y^2)Q_W^e+1\right]\,\eta_{\mu\nu}\,U_\kappa U_\lambda\Big\}$$
$$-\frac{e^4(Q_W^e+1)\,\mathrm{Re}\,(k_{\phi\phi})^{\mu\nu}}{2M_Z^2\,y\,(1-y)\sin^2 2\theta_W}\Big\{(Q_W^e-1)\,\left[(1-2y+2y^2)\,S_\mu S_\nu-T_\mu T_\nu-U_\mu U_\nu\right]$$
$$+\,s\left[(1-y+y^2)Q_W^e+1+y-y^2\right]\,\eta_{\mu\nu}\Big\}\,, \tag{9}$$

and

$$\delta\overline{|\mathcal{M}_L|^2} = \frac{2e^4(k_W)^{\kappa\mu\lambda\nu}}{s\,M_Z^2\,y^2(1-y)^2\cos^2\theta_W}\Big\{(y-1)\left[(2-4y+y^2)\,Q_W^e-1+2y\right]\,S_\kappa T_\mu S_\lambda T_\nu$$
$$+\,y\,\left[(1-2y-y^2)Q_W^e-1+2y\right]\,S_\kappa U_\mu S_\lambda U_\nu+\left[(2-y+y^2)Q_W^e-1\right]\,T_\kappa U_\mu T_\lambda U_\nu$$
$$-\,s\,y\,(1-y)\,[y\,(2-y)Q_W^e-1]\,\eta_{\mu\nu}\,T_\kappa T_\lambda-s\,y\,(1-y)\,\left[(1-y^2)Q_W^e-1\right]\,\eta_{\mu\nu}\,U_\kappa U_\lambda\Big\}$$
$$-\frac{e^4(Q_W^e-1)\,\mathrm{Re}\,(k_{\phi\phi})^{\mu\nu}}{2M_Z^2\,y\,(1-y)\sin^2 2\theta_W}\Big\{(Q_W^e+1)\,\left[(1-2y+2y^2)\,S_\mu S_\nu-T_\mu T_\nu-U_\mu U_\nu\right]$$
$$+\,s\left[(1-y+y^2)Q_W^e-1-y+y^2\right]\,\eta_{\mu\nu}\Big\}\,. \tag{10}$$

In these expressions, $s=S^2=(k+p)^2$ denotes the usual Mandelstam variable corresponding to the center-of-mass energy of the system. Employing the notation of Ref. [29], we have defined $y\equiv -s^{-1}T^2=-s^{-1}(k'-k)^2$, which provides a measure of the scattering angle. In the above results, terms of order M_Z^{-3} and higher have been dropped, and the ultrarelativistic limit for the external electron momenta has been adopted by setting explicitly appearing m to zero. Equations (9) and (10) are the primary theoretical result of this report. We remark that the last lines of both Eqs. (9) and (10) contain the Lorentz-symmetric trace part of $k_{\phi\phi}$, which is independently unobservable and can thus be disregarded.

Actual measurements of $\overline{|\mathcal{M}_R|^2}$ and $\overline{|\mathcal{M}_L|^2}$ are often combined into the following asymmetry observable:

$$A\equiv\frac{d\sigma_R-d\sigma_L}{d\sigma_R+d\sigma_L}=\frac{\overline{|\mathcal{M}_R|^2}-\overline{|\mathcal{M}_L|^2}}{\overline{|\mathcal{M}_R|^2}+\overline{|\mathcal{M}_L|^2}}\,. \tag{11}$$

Just as the individual cross sections (5), the asymmetry $A=A_0+\delta A$ consists of a Lorentz-invariant contribution [29]

$$A_0=\frac{G_F}{2\sqrt{2}\pi\alpha}\,\frac{y\,(1-y)}{(y^2-y+1)^2}\,s\,Q_W^e\,. \tag{12}$$

and a Lorentz-breaking correction [26]

$$\delta A=\frac{G_F}{\sqrt{2}\pi\alpha}\,\frac{(k_W)^{\kappa\mu\lambda\nu}}{(y^2-y+1)^2}\,\frac{\sin^2\theta_W}{s}\,\big[(1-2y)\,y\,S_\kappa U_\mu S_\lambda U_\nu-(2y^2-3y+1)\,S_\kappa T_\mu S_\lambda T_\nu$$
$$+\,T_\kappa U_\mu T_\lambda U_\nu-s\,(1-y)\,y\,\eta_{\mu\nu}\,(T_\kappa T_\lambda+U_\kappa U_\lambda)\big]\,. \tag{13}$$

IOP Conf. Series: Journal of Physics: Conf. Series **952** (2018) 012008 doi:10.1088/1742-6596/952/1/012008

These expressions are again correct at leading order in M_Z^{-2} and in the ultrarelativistic limit for the external momenta. For convenience, we have expressed M_Z in terms of the fine-structure constant α and the Fermi constant G_F. It turns out that the asymmetry correction δA depends on $k_{\phi\phi}$ only through the unobservable Lorentz-invariant trace of $k_{\phi\phi}$, which has been dropped from Eq. (13).

3. Experimental tests

In principle, the general mSME prediction (13) for the dominant Lorentz-breaking effects in Møller scattering allows the extraction of experimental constraints on the full k_W. However, in what follows we assume a special structure for k_W for simplicity:

$$(k_W)^{\mu\nu\rho\sigma} = \tfrac{1}{2}\left[\eta^{\mu\rho}\zeta^{\{\nu}\xi^{\sigma\}} - \eta^{\mu\sigma}\zeta^{\{\nu}\xi^{\rho\}} + \eta^{\nu\sigma}\zeta^{\{\mu}\xi^{\rho\}} - \eta^{\nu\rho}\zeta^{\{\mu}\xi^{\sigma\}}\right], \tag{14}$$

where

$$\zeta^{\{\mu}\xi^{\nu\}} \equiv \tfrac{1}{2}(\zeta^\mu\xi^\nu + \zeta^\nu\xi^\mu), \quad \zeta^\mu = (1,\vec{0}), \quad \xi^\mu = (0,\vec{\xi}). \tag{15}$$

This assumption reduces the number of Lorentz-breaking coefficients from 19 for the general k_W to the three components of $\vec{\xi}$. Note that the analogous three coefficients in the photon sector are constrained by the weakest experimental limits and have thus also been investigated separately as a special case [30, 13]. With this assumption, δA simplifies to

$$\delta A = \frac{G_F}{\sqrt{2}\pi\alpha}\frac{\zeta^{\{\mu}\xi^{\nu\}}\sin^2\theta_W}{(y^2-y+1)^2}\left[k'_\mu k'_\nu + y\,p_\mu p_\nu + (1-y)\,k_\mu k_\nu - 2(1-y)\,k_\mu k'_\nu - 2y\,p_\mu k'_\nu\right]. \tag{16}$$

Here, we have reverted to the original momentum variables with p'_μ eliminated by energy–momentum conservation, and we have again implemented the ultrarelativistic limit $m \to 0$.

In the mSME, in flat spacetime, the Lorentz-breaking k_W, and hence ζ and ξ, are usually considered to be position independent. Their components in cartesian inertial coordinates are then constant. The standard frame in which to express these and other SME coefficients is the Sun-centered celestial equatorial frame (SCCEF) [8]. It is therefore natural (but not necessary) to continue our analysis in the SCCEF. To this end, we need to transform all momenta appearing in Eq. (16) from the terrestrial laboratory frame, where they are typically measured, to the SCCEF. The dominant motion of the laboratory with respect to the SCCEF is the rotation of the Earth about its axis. Other effects, such as the associated boosts or the motion around the Sun can be neglected for our present purposes.

The explicit transformation between laboratory and Sun-centered frames requires a definite laboratory frame. Our choice for a laboratory at colatitude χ is characterized by an xy-plane parallel to the local surface of the Earth with the x-axis pointing South and the y-axis East. Right-handed cartesian coordinates then have the z-axis pointing vertically upward. This yields the following rotation matrix converting between the laboratory frame and the SCCEF [10]:

$$R^{Jj}(t) = \begin{pmatrix} \cos\chi\cos\Omega_\oplus t & -\sin\Omega_\oplus t & \sin\chi\cos\Omega_\oplus t \\ \cos\chi\sin\Omega_\oplus t & \cos\Omega_\oplus t & \sin\chi\sin\Omega_\oplus t \\ -\sin\chi & 0 & \cos\chi \end{pmatrix}. \tag{17}$$

Here, $\Omega_\oplus = 2\pi/(23\,\mathrm{h}\ 56\,\mathrm{min})$ is the Earth's sidereal angular frequency, $J = X, Y, Z$ are the spatial SCCEF components, and $j = x, y, z$ the spatial laboratory-frame components. The time dependence of $R^{Jj}(t)$ implies that the spatial components of various momenta that appear constant in the laboratory frame are in general time dependent in the SCCEF. In particular, SCCEF momenta for the beam electrons, the target electrons, and the collected scattered electrons take the form $k^\mu = (E_k, \vec{k}(t))$, $p^\mu = (m, \vec{0})$, and $k'^\mu = (E_{k'}, \vec{k}'(t))$, respectively.

Expressing the magnitude of all 3-momenta in terms of the beam energy E_k and the previously introduced variable y,[1] and considering an incoming-beam direction that is parallel to the local surface of the Earth and points in a direction α East of South, we find

$$\begin{aligned}\delta A(t) &= \frac{G_F}{\sqrt{2}\pi\alpha}\,\frac{E_k\,y\,(1-y)\sin^2\theta_W}{(y^2-y+1)^2}\,\vec{k}(t)\cdot\vec{\xi}\\ &= \frac{G_F}{\sqrt{2}\pi\alpha}\,\frac{E_k^2\,y\,(1-y)\sin^2\theta_W}{(y^2-y+1)^2}\times\\ &\quad\left[\sqrt{\xi_X^2+\xi_Y^2}\sqrt{1-\cos^2\alpha\sin^2\chi}\,\cos\Omega_\oplus t + c_0\right],\end{aligned} \tag{18}$$

where we have implemented the ultrarelativistic limit, as before. We have also absorbed an irrelevant phase into the definition of the origin of the time variable t. Moreover, we have dropped a constant shift in the asymmetry correction because it is difficult to disentangle from Lorentz-invariant effects. Our primary result is the square-root term; it predicts sidereal oscillation of the asymmetry δA with an amplitude determined by the Lorentz-breaking coefficients ξ_X and ξ_Y.

A measurement of this kind was carried out by the E158 experiment at SLAC [27]. The colatitude of this laboratory is $\chi = 53°$, and the incoming-beam direction points $\alpha = 123°$ East of South. Incoming polarized electrons with an energy of roughly $E_k = 50\,\text{GeV}$ were scattered off the atomic electrons of a stationary liquid-hydrogen target. The outgoing Møller electrons were detected and counted for angles $\frac{1}{2} < y < \frac{3}{4}$. This data has permitted a measurement of the asymmetry A with a statistical uncertainty of 1.4×10^{-8}.

This result may be used to estimate possible limits on mSME coefficients ξ_X and ξ_Y that may be extracted from the full E158 data set. We begin with the assumption that the Lorentz-violating amplitude

$$a_\oplus \equiv \frac{\sqrt{1-\cos^2\alpha\sin^2\chi}\,E_k^2\,y\,(1-y)}{2\,M_Z^2\,(y^2-y+1)^2\cos^2\theta_W}\sqrt{\xi_X^2+\xi_Y^2} \tag{19}$$

of the predicted sidereal oscillations cannot be much larger than the statistical uncertainty of the measurement $a_\oplus \lesssim 1.4 \times 10^{-8}$. This yields

$$\sqrt{\xi_X^2+\xi_Y^2} \lesssim 3.4 \times 10^{-7}\,. \tag{20}$$

To obtain this estimate, the following additional consideration has been implemented: from the range $\frac{1}{2} < y < \frac{3}{4}$ in the collected data, we have taken $y = \frac{3}{4}$. At this value, the prediction for $a_\oplus$ is smallest giving the most conservative estimate. Relative to previous studies preformed in other systems and with different methods, the result (20) represents an improvement by two orders of magnitude.

Acknowledgments

This report was supported in part by the Indiana University Center for Spacetime Symmetries under an IUCRG grant, by the Alexander von Humboldt Foundation, and by the Fundação para a Ciência e a Tecnologia of Portugal (FCT) through Project No. UID/FIS/00099/2013.

References

[1] Kostelecký V A and Samuel S 1989 *Phys. Rev.* D **39** 683
Kostelecký V A and Samuel S 1991 *Nucl. Phys.* B **359** 545

[1] The quantity y is understood to be entirely different from the coordinate component y.

Hashimoto K and Murata M 2013 *PTEP* **2013** 043B01
[2] Mocioiu I, Pospelov M and Roiban R 2000 *Phys. Lett.* B **489** 390
Carroll S M, Harvey J A, Kostelecký V A, Lane C D and Okamoto T 2001 *Phys. Rev. Lett.* **87** 141601
Carlson C E, Carone C D and Lebed R F 2001 *Phys. Lett.* B **518** 201
Anisimov A, Banks T, Dine M and Graesser M 2002 *Phys. Rev.* D **65** 085032
[3] Kostelecký V A, Lehnert R and Perry M J 2003 *Phys. Rev.* D **68** 123511
Jackiw R and Pi S-Y 2003 *Phys. Rev.* D **68** 104012
Bertolami O, Lehnert R, Potting R and Ribeiro A 2004 *Phys. Rev.* D **69** 083513
[4] Alfaro J, Morales-Técotl H A and Urrutia L F 2000 *Phys. Rev. Lett.* **84** 2318
Klinkhamer F R and Rupp C 2004 *Phys. Rev.* D **70** 045020
Amelino-Camelia G, Lämmerzahl C, Macias A and Müller H 2005 *AIP Conf. Proc.* **758** 30
Mavromatos N E 2005 *Lect. Notes Phys.* **669** 245
[5] Kostelecký V A and Samuel S 1990 *Phys. Rev.* D **42** 1289
Arkani-Hamed N, Cheng H C, Luty M A and Mukohyama S 2004 *J. High Energy Physics* JHEP05(2004)074
Libanov M V and Rubakov V A 2005 *J. High Energy Physics* JHEP08(2005)001
Dvali G, Pujolas O and Redi M 2007 *Phys. Rev.* D **76** 044028
Dubovsky S, Tinyakov P and Zaldarriaga M 2007 *J. High Energy Physics* JHEP11(2007)083
[6] Colladay D and Kostelecký V A 1997 *Phys. Rev.* D **55** 6760
Colladay D and Kostelecký V A 1998 *Phys. Rev.* D **58** 116002
[7] Kostelecký V A 2004 *Phys. Rev.* D **69** 105009
[8] For an overview of various tests of Lorentz symmetry see Kostelecký V A and Russell N 2011 *Rev. Mod. Phys.* **83** 11 [2017 edition: arXiv:0801.0287v10]
[9] See, e.g., Coleman S R and Glashow S L 1997 *Phys. Lett.* B **405** 249
Jacobson T, Liberati S and Mattingly D 2002 *Phys. Rev.* D **66** 081302
Lehnert R 2003 *Phys. Rev.* D **68** 085003
Klinkhamer F R and Schreck M 2008 *Phys. Rev.* D **78** 085026
Scully S T and Stecker F W 2009 *Astropart. Phys.* **31** 220
[10] Kostelecký V A and Mewes M 2002 *Phys. Rev.* D **66** 056005
[11] Lehnert R and Potting R 2004 *Phys. Rev. Lett.* **93** 110402
Lehnert R and Potting R 2004 *Phys. Rev.* D **70** 125010
Kostelecký V A and Mewes M 2009 *Phys. Rev.* D **80** 015020
Abdo A A *et al.* 2009 *Science* **323** 1688
[12] Kostelecký V A and Potting R 1995 *Phys. Rev.* D **51** 3923
Hohensee M A, Lehnert R, Phillips D F and Walsworth R L 2009 *Phys. Rev. Lett.* **102** 170402
Hohensee M A, Lehnert R, Phillips D F and Walsworth R L 2009 *Phys. Rev.* D **80** 036010
Altschul B 2009 *Phys. Rev.* D **80** 091901
Amelino-Camelia G *et al.* 2010 *Eur. Phys. J.* C **68** 619
Di Domenico A *et al.* 2010 Found. Phys. **40** 852
Berger M S, Kostelecký V A and Liu Z 2016 *Phys. Rev.* D **93** 036005
[13] Bocquet J-P *et al.* 2010 *Phys. Rev. Lett.* **104** 241601
[14] Barger V, Marfatia D, and Whisnant K 2007 *Phys. Lett.* B **653** 267
Adamson P *et al.* 2010 *Phys. Rev. Lett.* **105** 151601
Cohen A G and Glashow S L 2011 *Phys. Rev. Lett.* **107** 181803
Kostelecký V A and Mewes M 2012 *Phys. Rev.* D **85** 096005
Katori T 2012 *Mod. Phys. Lett.* A **27** 1230024
Díaz J S, Kostelecký V A and Lehnert R 2013 *Phys. Rev.* D **88** 071902
[15] Kostelecký V A and Lehnert R 2001 *Phys. Rev.* D **63** 065008
Adam C and Klinkhamer F R 2003 *Nucl. Phys.* B **657** 214
Belich Jr. H, Costa-Soares T, Ferreira Jr. M M and Helayël-Neto J H 2005 *Eur. Phys. J.* C **42** 127
Reyes C M Urrutia L F and Vergara J D 2008 *Phys. Rev.* D **78** 125011
Colladay D McDonald P and Potting R 2014 *Phys. Rev.* D **89** 085014
Hernaski C 2014 *Phys. Rev.* D **90** 124036
[16] Kostelecký V A Lane C D and Pickering A G M 2002 *Phys. Rev.* D **65** 056006
de Berredo-Peixoto G and Shapiro I L 2006 *Phys. Lett.* B **642** 153
Colladay D and McDonald P 2009 *Phys. Rev.* D **79** 125019
Colladay D and McDonald P 2008 *Phys. Rev.* D **77** 085006
Colladay D and McDonald P 2007 *Phys. Rev.* D **75** 105002
Ferrero A and Altschul B 2001 *Phys. Rev.* D **84** 065030
[17] Colladay D and McDonald P 2002 *J. Math. Phys.* **43** 3554

Lehnert R 2006 *Phys. Rev.* D **74** 125001
Lehnert R 2010 *Rev. Mex. Fis.* **56** 469
Altschul B 2006 *J. Phys.* A **39** 13757
Drummond I T 2013 *Phys. Rev.* D **88** 025009
Drummond I T 2017 *Phys. Rev.* D **95** 025006

[18] Jackiw R and Kostelecký V A 1999 *Phys. Rev. Lett.* **82** 3572
Pérez-Victoria M 1999 *Phys. Rev. Lett.* **83** 2518

[19] Belich Jr. H, Ferreira Jr. M M, Helayël-Neto J A and Orlando M T D 2003 *Phys. Rev.* D **67** 125011
Lehnert R 2004 *J. Math. Phys.* **45** 3399
Altschul B 2006 *Phys. Rev.* D **73** 036005
Bailey Q C and Kostelecký V A 2006 *Phys. Rev.* D **74** 045001
Hariton A J and Lehnert R 2007 *Phys. Lett.* A **367** 11
Gomes M, Nascimento J R, Petrov A Y and da Silva A J 2010 *Phys. Rev.* D **81** 045018
Kostelecký V A and Tasson J D 2011 *Phys. Rev.* D **83** 016013
Potting R 2012 *Phys. Rev.* D **85** 045033
Cambiaso M, Lehnert R and Potting R 2012 *Phys. Rev.* D **85** 085023
Cambiaso M, Lehnert R and Potting R 2014 *Phys. Rev.* D **90** 065003
Schreck M 2014 *Phys. Rev.* D **89** 105019
Schreck M 2014 *Phys. Rev.* D **90** 085025
Maluf R V, Silva J E G and Almeida C A S 2015 *Phys. Lett.* B **749** 304

[20] Kostelecký V A and Russell N 2010 *Phys. Lett.* B **693** 443
Kostelecký V A 2011 *Phys. Lett.* B **701** 137
Colladay D and McDonald P 2012 *Phys. Rev.* D **85** 044042
Kostelecký V A, Russell N and Tso R 2012 *Phys. Lett.* B **716** 470
Silva J E G and Almeida C A S 2014 *Phys. Lett.* B **731** 74
Schreck M 2015 *Eur. Phys. J.* C **75** 187
Schreck M 2015 *Phys. Rev.* D **91** 105001
Schreck M 2015 *Phys. Rev.* D **92** 125032
Russell N 2015 *Phys. Rev.* D **91** 045008
Foster J and Lehnert R 2015 *Phys. Lett.* B **746** 164
Colladay D and McDonald P 2015 *Phys. Rev.* D **92** 085031

[21] Kostelecký V A, Russell N and Tasson J D 2008 *Phys. Rev. Lett.* **100** 111102
Lehnert R, Snow W M and Yan H 2014 *Phys. Lett.* B **730** 353 [Erratum: 2015 *Phys. Lett.* B **744** 415]

[22] Foster J, Kostelecký V A and Xu R 2017 *Phys. Rev.* D **95** 084033
Lehnert R, Snow W M, Xiao Z and Xu R 2017 *Phys. Lett.* B **772** 865

[23] Iltan E O 2005 *Eur. Phys. J.* C **40** 269
Iltan E O 2004 *Mod. Phys. Lett.* A **19** 327
Iltan E O 2005 *Acta Phys. Polon.* B **36** 2955
Aranda J I, Ramírez-Zavaleta F, Rosete D A, Tlachino F J, Toscano J J and Tututi E S 2014 *J. Phys.* G **41** 055003
Aranda J I, Ramírez-Zavaleta F, Tlachino F J, Toscano J J and Tututi E S 2014 *Int. J. Mod. Phys.* A **29** 1450180

[24] Anderson D L, Sher M and Turan I 2004 *Phys. Rev.* D **70** 016001
Noordmans J P, Wilschut H W and Timmermans R G E 2013 *Phys. Rev.* C **87** 055502
Noordmans J P, Wilschut H W and Timmermans R G E 2013 *Phys. Rev. Lett.* **111** 171601
Dijck E A, Müller S E, Noordmans J P, Onderwater C J G, Pijpker C, Sytema A, Timmermans R G E, Vos K K, Wilschut H W 2013 *Annalen Phys.* **525** 653
Müller S E *et al.* 2013 *Phys. Rev.* D **88** 071901
Vos K K, Noordmans J P, Wilschut H W and Timmermans R G E 2014 *Phys. Lett.* B **729** 112
Noordmans J P and Vos K K 2014 *Phys. Rev.* D **89** 101702
Vos K K, Wilschut H W and Timmermans R G E 2015 *Phys. Rev.* C **91** 038501
Vos K K, Wilschut H W and Timmermans R G E 2015 *Phys. Rev.* C **92**, 052501
Noordmans J P, Onderwater C J G, Wilschut H W and Timmermans R G E 2016 *Phys. Rev.* D **93** 116001

[25] Altschul B 2007 *Astropart. Phys.* **28** 380

[26] Fu H and Lehnert R 2016 *Phys. Lett.* B **762** 33

[27] Anthony P L *et al.* [SLAC E158 Collaboration] 2004 *Phys. Rev. Lett.* **92** 181602
Anthony P L *et al.* [SLAC E158 Collaboration] 2005 *Phys. Rev. Lett.* **95** 081601

[28] Benesch J *et al.* [MOLLER Collaboration] 2014 The MOLLER Experiment: An Ultra-Precise Measurement of the Weak Mixing Angle Using Møller Scattering *Preprint* arXiv:1411.4088 [nucl-ex]

IOP Conf. Series: Journal of Physics: Conf. Series **952** (2018) 012008 doi:10.1088/1742-6596/952/1/012008

[29] See, e.g., Derman E and Marciano W J 1979 *Annals Phys.* **121** 147
[30] Schreck M 2012 *Phys. Rev.* D **86** 065038

IOP Conf. Series: Journal of Physics: Conf. Series **952** (2018) 012009 doi:10.1088/1742-6596/952/1/012009

Loop Corrections in Very Special Relativity Standard Model

Jorge Alfaro

Facultad de Física, Pontificia Universidad Católica de Chile, Casilla 306, Santiago 22, Chile.

E-mail: jalfaro@uc.cl

Abstract. In this talk we want to study one-loop corrections in VSRSM. In particular, we use the new $Sim(2)$-invariant dimensional regularization to compute one-loop corrections to the Effective Action in the subsector of the VSRSM that describe the interaction of photons with charged leptons. New stringent bounds for the masses of ν_e and ν_μ are obtained.

1. Introduction

The Standard Model of Particle Physics (SM) is being corroborated at the LHC at a great precision. Moreover, aside from the Higgs particle that was already a part of the SM, neither new particles nor new interactions have been found at the LHC till now [1].

But we know that the SM fails in the description of neutrinos. In the SM neutrinos are massless whereas in Nature at least some of the species of neutrinos must be massive because they exhibit neutrino oscillations [2].

To describe massive neutrinos in a Lorentz-invariant Quantum Field Theory (QFT), new kinds of particles must be postulated, as in the Seesaw Mechanism [3].

If we want to keep the particles and symmetries of the SM and give masses to the neutrinos we are led to Very Special Relativity(VSR) [4]. VSR postulates that the true symmetry of Nature is not the Lorentz group (6 parameters) but a subgroup of it, $Sim(2)$ (4 parameters). Doing so, none of the classical tests of Special Relativity (SR) is affected but it is possible to provide a non-local mass term for the neutrino [5].

Recently, we have used VSR to build the VSRSM [6]. It contains the same particles and symmetries as the SM, but a non-local term provides for the masses as well as neutrino oscillations. New predictions are obtained such as the process $\mu \to e + \gamma$, which is forbidden in the SM.

In this talk we want to consider loop corrections in the VSRSM. To do that we need to introduce a $Sim(2)$-invariant dimensional regularization [7], based on a simplification of the Mandelstam-Leibbrandt prescription [8]. Moreover, we choose to compute one loop corrections in the sector of the VSRSM that describe the coupling of photons to charge particles (VSRQED). Since the neutrino and the charged lepton of the same family form a doublet under $SU(2)_L$, they have a common VSR mass. Therefore, by computing processes in VSRQED, we can get information about neutrino masses.

2. The model

In this talk we restrict ourselves to the electron family. m is the VSR mass of both the electron and the neutrino, producing the second term of (1). After spontaneous symmetry breaking (SSB), the electron acquires a mass $M = \frac{G_e v}{\sqrt{2}}$, where G_e is the electron Yukawa coupling and v is the VEV of the Higgs, i.e. third term of (1). Please see equation (52) of [6]. The neutrino mass is not affected by SSB: $M_{\nu_e} = m$.

Restricting the VSRSM after SSB to the interactions between photon and electron alone, we get the VSRQED action:

$$\mathcal{L} = \bar{\psi}\left(i\left(\not{D} + \frac{1}{2}\not{n} m^2 (n \cdot D)^{-1}\right) - M\right)\psi - \frac{1}{4}F_{\mu\nu}F^{\mu\nu} - \frac{(\partial_\mu A^\mu)^2}{4} \tag{1}$$

where we denote the electron field by ψ, the photon field by A_μ and used the Feynman gauge. In (1) n_μ is a fixed null vector, i.e., $n \cdot n = 0$, while $D_\mu = \partial_\mu - ieA_\mu$, $F_{\mu\nu} = \partial_\mu A_\nu - \partial_\nu A_\mu$. We see that the electron mass, determined by the pole of the propagator, is $M_e = \sqrt{M^2 + m^2}$.

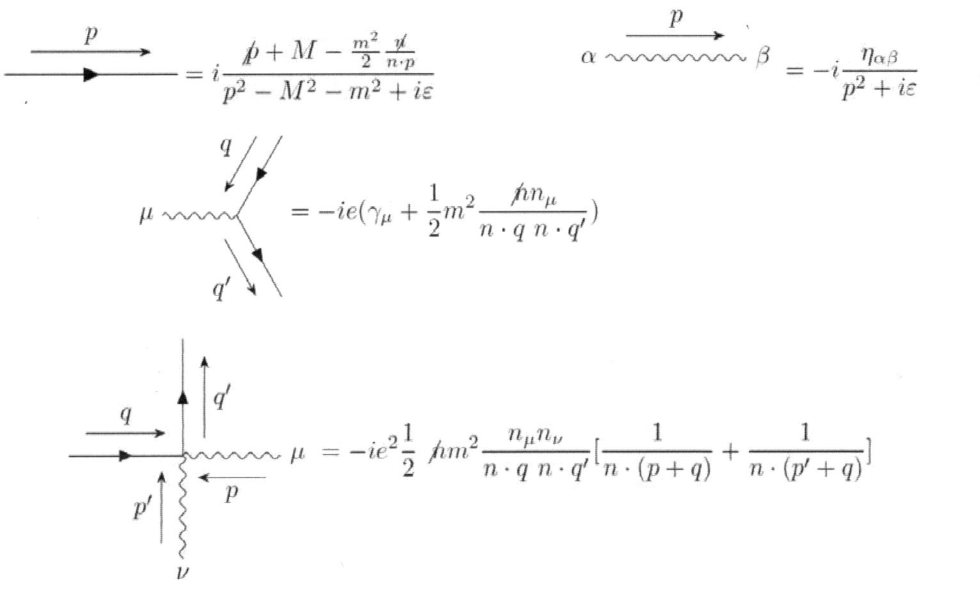

Figure 1. Feynman rules for one-loop computations: the electron and photon propagators, the $A_\mu ee$ and $A_\mu A_\nu ee$ vertices.

To draw the Feynman graphs we used [9]. In the following sections, we have used extensively the program FORM [10].

3. Photon Self Energy in VSRQED

In this section we present the computation of the photon self-energy. Applying the $Sim(2)$-invariant regulator to the addition of the graphs of Figure 2, we get:

$$i\Pi_{\mu\nu} = A(\eta_{\mu\nu}q^2 - q_\mu q_\nu) + B\left(-q^2\frac{n_\mu n_\nu}{(n \cdot q)^2} + \frac{n_\mu q_\nu + n_\nu q_\mu}{n \cdot q} - \eta_{\mu\nu}\right) \tag{2}$$

IOP Conf. Series: Journal of Physics: Conf. Series **952** (2018) 012009 doi:10.1088/1742-6596/952/1/012009

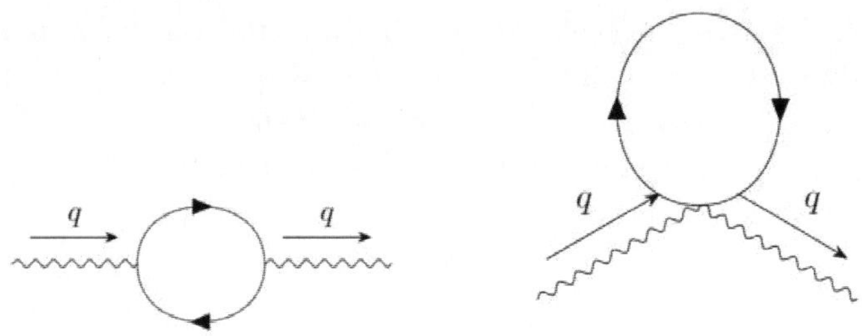

Figure 2. Vacuum polarization one-loop graphs

with

$$\begin{aligned} A &= (-ie)^2 \frac{i}{(4\pi)^\omega} \int_0^1 dx\, \Gamma(2-\omega) \frac{8x(1-x)}{(M_e^2 - (1-x)xq^2)^{2-\omega}} \\ B &= -m^2 i \frac{e^2}{4\pi^2} \int_0^1 \frac{dx}{(1-x)} \log\left[1 - \frac{q^2(1-x)^2}{M_e^2 - q^2(1-x)x}\right] . \end{aligned} \tag{3}$$

Here $-e$ is the electron electric charge, m the electron-neutrino mass, M_e the electron mass and q_μ is the virtual photon momentum.

We first notice that $q^\mu \Pi_{\mu\nu} = 0$ as required by $U(1)$ gauge invariance of the photon field. It is obtained by a straightforward application of the regularized integrals of [7]. Moreover $B(q^2 = 0) = 0$, therefore the photon remains massless. Also the photon wave-function divergence is the same as in QED.

4. Electron Self Energy in VSRQED

Here we calculate the electron self-energy. Again we have two graphs contributing to the 2-proper vertex, see Figure 3.

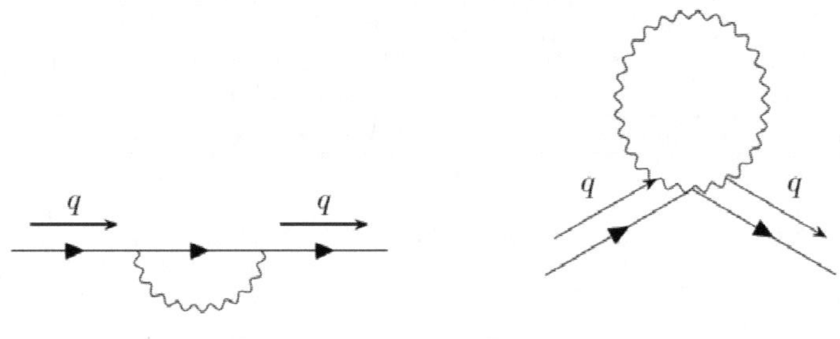

Figure 3. Electron self-energy one-loop graphs. The second graph vanishes in Feynman gauge.

It follows that

$$-i\Sigma(q) = C\frac{\not{n}}{n \cdot q} + D\not{q} + E \tag{4}$$

with:

$$\begin{aligned} C &= (-ie)^2 m^2 \Bigg[\frac{i}{16\pi^2} \int_0^1 dx (1-x)^{-1} \ln\left(1 + \frac{q^2(1-x)}{(M_e^2 - q^2 - i\varepsilon)}\right) \\ &\quad + 2i(4\pi)^{-\omega} \int_0^1 dx \frac{\Gamma(2-\omega)}{[\mu^2 x - x(1-x)q^2 + M_e^2(1-x) - i\varepsilon]^{2-\omega}} \Bigg], \qquad (5) \\ D &= -2(-ie)^2(\omega-1) i (4\pi)^{-\omega} \int_0^1 dx \frac{\Gamma(2-\omega)x}{[\mu^2 x - x(1-x)q^2 + M_e^2(1-x) - i\varepsilon]^{2-\omega}}, \qquad (6) \\ E &= (-ie)^2 2\omega M i (4\pi)^{-\omega} \int_0^1 dx \frac{\Gamma(2-\omega)}{[\mu^2 x - x(1-x)q^2 + M_e^2(1-x) - i\varepsilon]^{2-\omega}}. \qquad (7) \end{aligned}$$

5. Electron-Electron-Photon Proper vertex (EEP vertex) $\Gamma^\mu(p+q,p)$

In this section we discuss the EEP vertex and verify the Ward-Takahashi identity. This is an important test of the gauge invariance of the regulator. The one-loop contribution to $\Gamma^\mu(p' = p+q, p)$ consists of the addition of three graphs (see Figure 4).

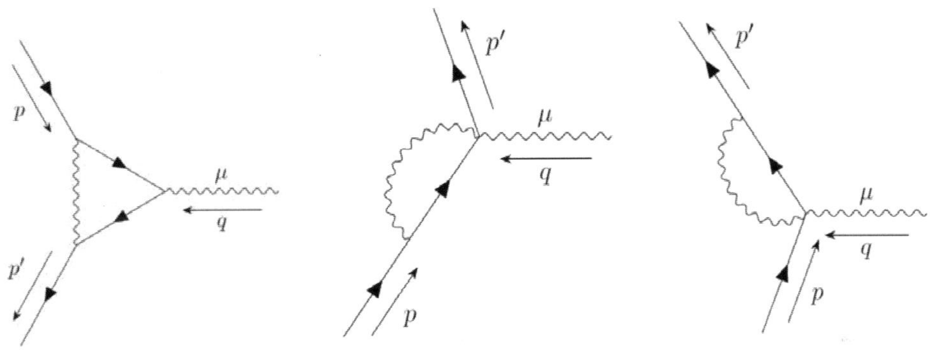

Figure 4. One-loop contributions to the EEP vertex

As a result of the shift symmetry which is respected by the regulator, $\int dp\, f(p_\mu) = \int dp\, f(p_\mu + q_\mu)$ for arbitrary q_μ, we can prove the Ward-Takahashi identity. This identity must hold for both the divergent and finite pieces of the Green functions:

$$-iq_\mu \Gamma^\mu(p+q,p) = S^{-1}(p+q) - S^{-1}(p) \qquad (8)$$

Here $S(p) = \frac{i}{\not{p} - M - \Sigma(p)}$ is the full electron propagator and $\Gamma^\mu(p+q,p)$ is the EEP vertex.

As a simple check, let us verify explicitly that the pole at $d = 4$ satisfies (8). For the pole contributions one finds:

$$\begin{aligned} P\Sigma(q) &= -(-ie)^2 \frac{1}{16\pi^2} \left\{ 2m^2 \frac{\not{n}}{n \cdot q} - \not{q} + 4M \right\} \frac{1}{2-\omega} \qquad (9) \\ P\Gamma^\mu(p+q,p) &= -(-ie)^2 \frac{1}{16\pi^2} \frac{1}{2-\omega} \left(\gamma_\mu + 2m^2 \not{n} \frac{n_\mu}{n\cdot p\; n\cdot(p+q)} \right) \qquad (10) \end{aligned}$$

The divergent piece satisfies the Ward-Takahashi identity (8):

$$q_\mu P\Gamma^\mu(p+q,p) = P\Sigma(p) - P\Sigma(p+q) = -(-ie)^2 \frac{1}{16\pi^2} \frac{1}{2-\omega} \left\{ \not{q} - 2m^2 \not{n} \left(\frac{1}{n\cdot p + n \cdot q} - \frac{1}{n \cdot p} \right) \right\}.$$

5.1. *Form factors*

The on-shell EEP vertex can be written as follows

$$M_\mu(p,q) = \bar{u}(p+q)\left\{G_2\left[-i\sigma_{\mu\nu}q_\nu\not{n}\right] + G_3\not{n}Q_\mu + F_3\not{n}\sigma_{\mu\nu}q_\nu\not{n} + \tilde{\gamma}_\mu F_1 + F_2 i\frac{\sigma_{\mu\nu}}{2M}q_\nu\right\}u(p) \quad (11)$$

where:

$$\tilde{\gamma}_\mu = \gamma_\mu + \frac{m^2}{2}\frac{\not{n}n_\mu}{n\cdot p\,(n\cdot p + n\cdot q)}, \qquad Q_\mu = q_\mu - q^2\frac{n_\mu}{n\cdot q}\,.$$

$F_1, F_2.F_3, G_2, G_3$ are forms factors (Lorentz-scalar combinations of n_μ, p_μ, q_μ). Under the $Sim(2)$ scaling $n_\mu \to \lambda n_\mu$, F_1, F_2 are invariants, $F_3 \to \lambda^{-2}F_3$, $G_3 \to \lambda^{-1}G_3$, $G_2 \to \lambda^{-1}G_2$.

In the non-relativistic (NR) limit we get Table 1, keeping terms that are at most linear in q_μ. The operator whose coefficient is the form factor, reduces in this limit to the addition of the terms appearing in the first column of Table 1, corresponding to the same form factor in the second column of Table 1.

Table 1. In the right column we list the form factor. In the left column we have the NR limit of the matrix element accompanying the form factor in (11). All form factors are evaluated at $q_\mu = 0$. Here A_0 is the electric potential and A_i is the vector potential. $\varphi_{s'}$ is a two-dimensional constant vector that corresponds to the NR limit of the Dirac spinors.

NR limit	Form factor
$2M_e\varphi_s^\dagger\varphi_s A_0$	$F_1(0)$
$\frac{3m^2}{4M^2}i\varepsilon_{ijk}\varphi_s^\dagger\sigma^i\varphi_{s'}\hat{n}_j q_k A_0$	$F_1(0)$
$i\varepsilon_{ijk}q_j\varphi_s^\dagger\sigma^k\varphi_{s'}A_i$	$F_1(0)$
$-2in_0M\varepsilon_{ijk}\varphi_s^\dagger\sigma^k\varphi_{s'}q_jA_i$	$G_2(0)$
$-i\varepsilon_{ijk}n_k\frac{m^2}{M}\varphi_s^\dagger\hat{n}.\vec{\sigma}\varphi_{s'}q_jA_i$	$G_2(0)$
$i(2M\varepsilon_{ijk}n_k\varphi_s^\dagger\sigma^j\varphi_{s'} + 2M_ein_i\varphi_s^\dagger\varphi_{s'})A_0q_i$	$G_2(0)$
$2M_en_0\varphi_s^\dagger\varphi_{s'}Q_\mu A^\mu$	$G_3(0)$
$(-4M_e\varepsilon_{ijk}n_k\varphi_s^\dagger\vec{n}.\vec{\sigma}\varphi_{s'} + 4M_en_0^2\varepsilon_{ijk}\varphi_s^\dagger\sigma^k\varphi_{s'})q_jA_i$	$F_3(0)$
$4M_en_0\varepsilon_{ijk}n_j\varphi_s^\dagger\sigma^k\varphi_{s'}A_0q_i$	$F_3(0)$
$i\varepsilon_{ijk}\varphi_s^\dagger\sigma^k\varphi_{s'}A_iq_j$	$F_2(0)$
$-i\frac{m^2}{2M^2}\varepsilon_{ijk}\hat{n}_j\varphi_s^\dagger\sigma^k\varphi_sA_0q_j$	$F_2(0)$

To show the power of the $Sim(2)$-invariant regularization prescription presented in [7], we will compute the one-loop contribution to the (isotropic) anomalous magnetic moment of the electron. It is given by $F_2(0) - 2n_0MG_2(0) - 4F_3(0)M_en_0^2i$ (see rows 11, 5 and 9 of Table 1).

Evaluating the integrals according to the $Sim(2)$-invariant prescription to $o(m^2)$, we get [7]:

$$F_2 - 4F_3M_en_0^2i - 2G_2n_0M = \frac{\alpha}{2\pi}$$

where α is the fine-structure constant. Therefore to this order the QED result holds.

Notice that already at tree level, the model predicts the existence of an anisotropic electric moment of the electron, corresponding to the second line of the list and an anisotropic magnetic moment of the electron, corresponding to the fourth row of the list, both of the order of $\frac{m^2}{M_e^2}$. The electric dipole moment is:

$$|\vec{p}| = \frac{3e}{4M_e}\frac{m^2}{M_e^2}|(\vec{s}\times\hat{n})|\frac{3}{8}\lambda e\frac{m^2}{M_e^2}$$

where $\lambda = 2.4 \times 10^{-12}$ m is the Compton wave length of the electron.

Using the best bound on the electric dipole moment of the electron [11], $|\vec{p}| < 8.7\times10^{-29}$ e·cm, we get:

$$\frac{m^2}{M_e^2} < 9.7 \times 10^{-19} .$$

For the muon $\lambda = 1.17 \times 10^{-14}$ m. Using the best bound on the muon electric dipole moment[12], $|\vec{p}_\mu| < 1.8 \times 10^{-19}$ e·cm, we get:

$$\frac{m_\mu^2}{M_\mu^2} < 4 \times 10^{-7} .$$

The $Sim(2)$-invariant regularization permits to explore the full quantum properties of VSR. They should be systematically tested, in Particle Physics models as well as in Quantum Gravity models.

Acknowledgments

J.A. wants to thank the organizers of the "Workshop on CPT and Lorentz Symmetry in Field theory" for a very stimulating atmosphere. The work of J.A. has been partially financed by Fondecyt 1150390 and Anillo ACT 1417.

References

[1] The CMS collaboration 2014 Evidence for the direct decay of the 125 GeV Higgs boson to fermions *Nature Physics* **10** 557–560
[2] Langacker P 2010 *The Standard model and Beyond* (Boca Raton, FL: CRC Press, Taylor & Francis)
[3] Mohapatra R 2002 *Unification and Supersymmetry: The Frontiers of Quark-Lepton Physics* 3rd ed (New York: Springer)
[4] Cohen A and Glashow S 2006 Very special relativity *Phys. Rev. Lett.* **97** 021601
[5] Cohen A and Glashow S 2006 A Lorentz-Violating Origin of Neutrino Mass? *Preprint* hep-ph/0605036
[6] Alfaro J, González P and Ávila R 2015 *Phys. Rev.* D **91** 105007 *Addendum: Phys. Rev.* D **91** 129904
[7] Alfaro J 2017 *Phys. Lett.* B **772** 100–104
[8] Alfaro J 2016 *Phys. Rev.* D **93** 065033 *Erratum Phys. Rev.* D **94** 049901
[9] Ellis J 2016 TikZ-Feynman: Feynman diagrams with TikZ *Preprint* hep-ph/1601.05437
[10] Vermaseren J 2000 New features of FORM *Preprint* math-ph/0010025
[11] The ACME Collaboration 2014 Order of Magnitude Smaller Limit on the Electric Dipole Moment of the Electron *Science* **343** (6168) 269–272
[12] Bennett G et al. (Muon (g-2) Collaboration) 2009 *Phys. Rev.* D **80** 052008

IOP Conf. Series: Journal of Physics: Conf. Series **952** (2018) 012010 doi:10.1088/1742-6596/952/1/012010

Casimir effect in presence of spontaneous Lorentz symmetry breaking

C A Escobar

CENTRA, Departamento de Física, Universidade do Algarve, 8005-139 Faro, Portugal

E-mail: cruiz@ualg.pt

Abstract. The Casimir effect is one of the most remarkable consequences of the nonzero vacuum energy predicted by quantum field theory. In this contribution we study the Lorentz-violation effects of the minimal standard-model extension on the Casimir force between two parallel conducting plates in the vacuum. Using a perturbative method, we compute the relevant Green's function which satisfies given boundary conditions. The standard point-splitting technique allow us to express the vacuum expectation value of the stress-energy tensor in terms of this Green's function. Finally, we study the Casimir energy and the Casimir force paying particular attention to the quantum effects as approaching the plates.

1. Introduction

Interest in Lorentz violation has grown rapidly in the last decades since many candidate theories of quantum gravity [1, 2], such as string theory [3, 4] and loop quantum gravity [5, 6, 7], possess scenarios involving deviations from Lorentz symmetry. Nowadays, investigations concerning Lorentz violation are mostly conducted under the framework of the standard model extension (SME), initiated by Kostelecký and Colladay [8, 9]. The SME is an effective field theory that contains the standard model, general relativity, and all possible operators that break Lorentz invariance. The Lorentz-violating (LV) coefficients arise as vacuum expectation values of some basic fields belonging to a more fundamental theory, such as string theory [10, 11]. Some important features of the minimal SME comprise invariance under observer Lorentz transformations, energy-momentum conservation, gauge invariance, power-counting renormalizability [12], causality, stability and hermiticity [13].

The main goal of this contribution is to provide additional results regarding the local effects of the quantum vacuum in a particular sector of the electrodynamics limit of the SME, namely, the CPT-odd Maxwell-Chern-Simons term [14]. Concretely, we study the Casimir effect (CE) between two parallel conducting plates using a local approach based on the calculation of the vacuum expectation value of the stress-energy tensor via Green's functions satisfying the suitable boundary conditions.

In its simple manifestation, the CE is a quantum force of attraction between two parallel uncharged conducting plates [15]. More generally, it refers to the stress on bounding surfaces when a quantum field is confined to a finite volume of space. The boundaries can be material media, interfaces between two phases of the vacuum, or topologies of space. In any case, the modes of the quantum fields are restricted, giving rise to a macroscopically measurable force [16]. The CE has been previously considered within the SME framework in Refs. [17, 18, 19].

The authors in Ref. [18] used the zeta-function-regularization technique to compute the Casimir force between two parallel conducting plates within the (3+1)D Maxwell-Chern-Simons theory. The first attempt to tackle this problem was due to M. Frank and I. Turan [17]; however, as pointed out by O. G. Kharlanov and V. Ch. Zhukovsky [18], they used misinterpreted equations which led to an oversimplified treatment of the problem. More precisely, they considered that the photon dispersion relation corresponds to that for a massive photon; however, unlike the (2+1)D case, in (3+1)D the effect of the Maxwell-Chern-Simons term is a more complicated dispersion relation for the photon. Due to this wrong equation, Frank and Turan constructed also incorrectly the relevant Green's function (GF). One of the specific aims of this contribution is the construction of the correct Green's function within the (3+1)D Maxwell-Chern-Simons theory and the calculation of the Casimir-energy density and stress between two parallel conducting plates.

The outline of this contribution is as follows. Section 2 reviews some basics of the particular sector of the minimal SME to be considered in this work, namely, the (3+1)D Maxwell-Chern-Simons model. Using a perturbative method similar to that used for obtaining the Born series for the scattering amplitudes in quantum mechanics, in section 3 we compute the leading-order Green's function which satisfies given Dirichlet, Neumann or Robin boundary conditions, provided the smallness of the LV coefficients. In section 4 we use the standard point-splitting technique to express the vacuum expectation value of the stress-energy tensor in terms of the Green's function. The concrete calculation of the renormalized vacuum stress (and the Casimir force) between two parallel conducting plates is performed in section 5. We also discuss the local energy density, which is found to diverge as approaching the plates. We demonstrate that the divergent term does not contribute to the observable force. We summarize our results in section 6. Details about this investigation were recently published in Ref. [20].

2. Lorentz-violating electrodynamics

In the present contribution we are concerned with the CPT-odd sector of the SME. The relevant Lagrangian is

$$\mathcal{L} = -\frac{1}{4} F_{\mu\nu} F^{\mu\nu} + (k_{AF})_\mu A_\nu \tilde{F}^{\mu\nu} - j_\mu A^\mu. \tag{1}$$

Here, $j^\mu = (\rho, \mathbf{J})$ is the 4-current source that couples to the electromagnetic 4-potential A^μ, $F^{\mu\nu} = \partial^\mu A^\nu - \partial^\nu A^\mu$ is the electromagnetic field strength and $\tilde{F}^{\mu\nu} = \frac{1}{2}\epsilon^{\mu\nu\alpha\beta} F_{\alpha\beta}$ its dual. From now on, we omit the subscript AF of the Lorentz- and CPT-violating $(k_{AF})^\mu$ coefficients and set $(k_{AF})^\mu \equiv k^\mu = (k^0, \mathbf{k})$. A nondynamical fixed k^μ determines a special direction in spacetime. For example, certain features of plane wave propagating along $\mathbf{k}$ might differ from those of waves perpendicular to $\mathbf{k}$. Thus, particle Lorentz transformations are violated.

Varying the action $\mathcal{S} = \int \mathcal{L}\, d^4x$ with respect to A^μ yields the equations of motion:

$$\left(\Box \eta^\mu{}_\nu - \partial^\mu \partial_\nu - 2k_\beta \epsilon^{\mu\beta\alpha}{}_\nu \partial_\alpha \right) A^\nu = j^\mu, \tag{2}$$

which extend the usual covariant Maxwell equations to incorporate Lorentz violation. Of course, the homogeneous Maxwell equations that express the field-potential relationship

$$\partial_\mu \tilde{F}^{\mu\nu} = 0, \tag{3}$$

are not modified due to the $U(1)$ gauge invariance of the action. The stress-energy tensor for this theory is, up to a total-derivative term, given by [9]

$$\Theta^{\mu\nu} = -F^{\mu\alpha} F^\nu{}_\alpha + \frac{1}{4} \eta^{\mu\nu} F^{\alpha\beta} F_{\alpha\beta} - k^\nu \tilde{F}^{\mu\alpha} A_\alpha. \tag{4}$$

Here $\eta^{\mu\nu} = \text{diag}(1,-1,-1,-1)$ denotes the usual Minkowski flat space-time metric. Unlike the conventional case, $\Theta^{\mu\nu}$ cannot be symmetrized because its antisymmetric part is not longer a total derivative. By virtue of the equations of motion (2) and (3), the energy-momentum tensor obeys

$$\partial_\mu \Theta^{\mu\nu} = j_\mu F^{\mu\nu}. \tag{5}$$

Although the energy-momentum tensor is gauge dependent, it only changes by a total derivative under the gauge transformation $A^\mu \to A^\mu + \partial^\mu \Lambda$, i.e.

$$\tilde{F}^{\mu\alpha} A_\alpha \to \tilde{F}^{\mu\alpha} A_\alpha - \partial_\alpha(\tilde{F}^{\mu\alpha}\Lambda). \tag{6}$$

Consequently, the integrals over all space are gauge invariant. Note that the energy

$$\mathcal{E} = \int \Theta^{00} d^3\mathbf{x} = \int \frac{1}{2}\left(\mathbf{E}^2 + \mathbf{B}^2 - k^0 \mathbf{B}\cdot\mathbf{A}\right) d^3\mathbf{x}, \tag{7}$$

is not positive definite due to the term $k^0\mathbf{B}\cdot\mathbf{A}$, which may be negative. The appearance of this term in the energy density can introduce instability in the theory [21, 22], and it can be resolved by requiring that only spacelike components of k^μ are nonzero. However, this condition depends on the observer frame, so even an infinitesimal boost to another observer frame would reintroduce instability. Despite arising from a hitherto unobserved spontaneous breaking of the electromagnetic U(1) gauge symmetry, the photon mass can be introduced in this theory to eliminate the linear instability [23]. Although this idea might be physically acceptable, in this work we restrict ourselves to the minimal extension with a purely spacelike background $k^\mu = (0, \mathbf{k})$, which is fundamentally different from the theory with the purely timelike case $k^\mu = (k^0, \mathbf{0})$, as reviewed extensively in Refs. [24, 25].

3. Green's function method

To derive the GF for the previously discussed LV electrodynamics one can employ standard Fourier methods. As in conventional electrodynamics, the modified Maxwell operator appearing in parentheses in Eq. (2) is singular. To circumvent the non invertibility of the corresponding Minkowski matrix one can further work in the Lorentz gauge. The free-space GF (satisfying the standard boundary conditions at infinity) in momentum [26, 27] and coordinate [24, 25] representations can be obtained in a simple fashion. In the present work we are concerned with the effects of this Lorentz-violating electrodynamics on the Casimir force between two parallel conducting plates in the vacuum. To this end we employ a local approach consisting in the evaluation of the vacuum expectation value of the stress-energy tensor of the system, which can be expressed in terms of the appropriate Green's function. The presence of boundaries (e.g. the plates) makes the GF derived in Refs. [24, 25, 26, 27] not suitable for our purposes. Thus the aim of this section is the construction of the Green's functions which incorporates the presence of boundaries.

In the Lorentz gauge $\partial_\mu A^\mu = 0$, the field equations (2) take the form

$$\left(\Box \eta^\mu{}_\nu - 2k_\beta \epsilon^{\mu\beta\alpha}{}_\nu \partial_\alpha\right) A^\nu = j^\mu, \tag{8}$$

where $\Box = \partial_\mu\partial^\mu = \partial_t^2 - \nabla^2$ is the D'Alembert operator. To obtain the general solution of Eq. (8) for arbitrary external sources, we introduce the GF matrix $G^\mu{}_\nu(x,x')$ solving Eq. (8) for a pointlike source,

$$\left(\Box \eta^\mu{}_\nu - 2k_\beta \epsilon^{\mu\beta\alpha}{}_\nu \partial_\alpha\right) G^\nu{}_\gamma(x,x') = \eta^\mu{}_\gamma \delta^4(x-x'), \tag{9}$$

in such a way that the general solution for the 4-potential in the Lorentz gauge is

$$A^{\mu}(x) = \int G^{\mu}{}_{\nu}(x,x')j^{\nu}(x')d^4x. \tag{10}$$

Since the timelike theory, without a photon-mass term, appears to be inconsistent (that is, the theory violates unitary and causality, or both), in this work we specialize to the purely spacelike case $k^{\mu} \equiv (0,\mathbf{k}) \equiv (0,0,0,\kappa)$. Without loss of generality, we consider surfaces Σ_i which are orthogonal to $\mathbf{k}$ in which Dirichlet, Neumann or Robin boundary conditions have been imposed. In this way, the GF we consider has translational invariance in the directions x and y, while this invariance is broken in the z-direction. Exploiting this symmetry we further introduce the Fourier transform in the direction parallel to the surfaces Σ_i, taking the coordinate dependence to be $\mathbf{R} = (x-x', y-y')$, and define

$$G^{\mu}{}_{\nu}(x,x') = \int \frac{d^2\mathbf{p}}{(2\pi)^2} e^{i\mathbf{p}\cdot(\mathbf{x}-\mathbf{x}')} \int \frac{d\omega}{2\pi} e^{-i\omega(t-t')} g^{\mu}{}_{\nu}(\omega,\mathbf{p},z,z'), \tag{11}$$

where $\mathbf{p} = (p_x, p_y)$ is the momentum parallel to Σ_i. Hereafter we will suppress the dependence of the reduced GF $g^{\mu}{}_{\nu}$ on ω and $\mathbf{p}$. The substitution of Eq. (11) into Eq. (9) yields the reduced GF equation

$$\left(\tilde{\Box}\eta^{\mu}{}_{\nu} - 2i\kappa\epsilon^{3\mu\beta}{}_{\nu}p_{\beta}\right) g^{\nu}{}_{\alpha}(z,z') = \eta^{\mu}{}_{\alpha}\delta(z-z'), \tag{12}$$

where $p^{\alpha} = (\omega, \mathbf{p}, 0)$ and $\tilde{\Box} = \mathbf{p}^2 - \omega^2 - \partial_z^2$. We now must solve the reduced GF equation for the various components. The solution to Eq. (12) is simple but not straightforward. Since the coefficient κ is assumed to be small, to solve it we employ a method similar to that used for obtaining the scattering amplitudes in quantum mechanics, in which the Schrödinger equation can be written as an integral equation, the Lippmann-Schwinger equation, which can be iterated to obtain the Born series. Indeed, the Lippmann-Schwinger equation for Green's operator is called the resolvent identity. In the problem at hand let us consider that the free (with $\kappa = 0$) reduced GF is known, being the solution of $\tilde{\Box}\mathfrak{g}(z,z') = \delta(z-z')$ in the region $\mathfrak{D} \subseteq \mathbb{R}$ and satisfying appropriate boundary conditions on the surfaces $\Sigma_i \subseteq \mathbb{R}^2$. Now Eq. (12) can be directly integrated using the free reduced GF. We thus establish the integral equation

$$g^{\mu}{}_{\nu}(z,z') = \eta^{\mu}{}_{\nu}\mathfrak{g}(z,z') + 2i\kappa\epsilon^{3\mu\alpha}{}_{\beta}p_{\alpha} \int_{\mathfrak{D}} \mathfrak{g}(z,z'')g^{\beta}{}_{\nu}(z'',z')dz''. \tag{13}$$

Suppose we take this expression for $g^{\beta}{}_{\nu}$, and plug it under the integral sign. Iterating this procedure, we obtain a formal series for $g^{\mu}{}_{\nu}$. At leading order in the LV coefficient κ, the reduced GF can be written as the sum of two terms, $g^{\mu}{}_{\nu}(z,z') = \eta^{\mu}{}_{\nu}\mathfrak{g}(z,z') + \mathbf{g}^{\mu}{}_{\nu}(z,z')$, where the first term provides the propagation in the absence of Lorentz violation, while the second term, to second order in the LV coefficient κ, encoding the Lorentz symmetry breakdown is given by

$$\mathbf{g}^{\mu}{}_{\nu}(z,z') = 2i\kappa\epsilon^{\mu}{}_{\nu}{}^{\alpha 3}p_{\alpha}N(z,z') - 4\kappa^2\left[p^{\mu}p_{\nu} - (\eta^{\mu}{}_{\nu} + n^{\mu}n_{\nu})p^2\right]M(z,z'), \tag{14}$$

where $n^{\mu} = (0,0,0,1)$ is the normal to the surfaces Σ_i. In deriving Eq. (14) we have used the identity $\epsilon^{3\mu\beta}{}_{\alpha}\epsilon^{3\alpha\gamma}{}_{\nu}p_{\beta}p_{\gamma} = p^{\mu}p_{\nu} - (\eta^{\mu}{}_{\nu} + n^{\mu}n_{\nu})p^2$, with the definitions

$$N(z,z') = \int_{\mathfrak{D}} \mathfrak{g}(z,z'')\mathfrak{g}(z'',z')dz'', \qquad M(z,z') = \int_{\mathfrak{D}} \mathfrak{g}(z,z'')N(z'',z')dz''. \tag{15}$$

IOP Conf. Series: Journal of Physics: Conf. Series **952** (2018) 012010 doi:10.1088/1742-6596/952/1/012010

To calculate the $M(z,z')$ and $N(z,z')$ functions we require the suitable reduced Green's functions, which are given by

$$\mathfrak{g}_0(z,z') = \frac{i}{2p}e^{ip(z_>-z_<)}, \qquad \mathfrak{g}_{||}(z,z') = \frac{\sin[pz_<]\sin[p(D-z_>)]}{p\sin[pD]}, \tag{16}$$

in the free space $\mathfrak{g}_0$, which is defined on the real line, i.e. $\mathcal{D} = \mathbb{R}$, and $\mathfrak{g}_{||}$ in presence of two parallel conducting plates separated by a distance D, which is defined on the domain $\mathcal{D}_{||} = [0,D]$ and satisfies the boundary conditions $\mathfrak{g}_{||}(0,z') = \mathfrak{g}_{||}(D,z') = 0$, respectively. In the above expressions $z_>$ ($z_<$) is the biggest (smaller) of z and z'.

Notice that the full GF matrix $G^{\mu}{}_{\nu}$ can be written as the sum of two terms, $G^{\mu}{}_{\nu}(x,x') = \eta^{\mu}{}_{\nu}\mathcal{G}(x,x') + \mathbf{G}^{\mu}{}_{\nu}(x,x')$, where $\mathcal{G}$ and $\mathbf{G}^{\mu}{}_{\nu}$ are the Fourier transformations of $\mathfrak{g}(z,z')$ and $\mathbf{g}^{\mu}{}_{\nu}$, respectively, as defined in Eq. (11). It is worth mentioning that the second term satisfies the Lorentz gauge condition, i.e. $\partial_\mu \mathbf{G}^{\mu}{}_{\nu} = 0$. The proof follows from the reduced GF: $\partial_\mu \mathbf{G}^{\mu}{}_{\nu} \propto \int p_\mu \mathbf{g}^{\mu}{}_{\nu}$, which vanishes given that $\epsilon^{\mu}{}_{\nu}{}^{\alpha 3} p_\mu p_\alpha = 0$ and $p_\mu n^\mu = 0$.

4. Vacuum stress-energy tensor

In section 2 we gave the stress-energy tensor (SET) for this theory and we showed that it can be written as the sum of two terms:

$$\Theta^{\mu\nu} = T^{\mu\nu} + \Xi^{\mu\nu}. \tag{17}$$

The first term,

$$T^{\mu\nu} = -F^{\mu\alpha}F^{\nu}{}_{\alpha} + \frac{1}{4}\eta^{\mu\nu}F^{\alpha\beta}F_{\alpha\beta}, \tag{18}$$

is the standard Maxwell stress-energy tensor, while the second,

$$\Xi^{\mu\nu} = -k^{\nu}\tilde{F}^{\mu\alpha}A_{\alpha}, \tag{19}$$

explicitly depends on the LV coefficients k^μ. Now we address the problem of the vacuum expectation value of the SET, to which we will refer simply as the vacuum stress (VS). This VS can be obtained from appropriate derivatives of the GF, in virtue of the formula

$$G^{\mu\nu}(x,x') = -i\langle 0|\hat{\mathcal{T}}A^{\mu}(x)A^{\nu}(x')|0\rangle. \tag{20}$$

Using the standard point splitting technique and taking the vacuum expectation value of the SET we can obtain

$$\langle\Theta^{\mu\nu}\rangle = \langle T^{\mu\nu}\rangle + \langle\Xi^{\mu\nu}\rangle, \tag{21}$$

where the first term,

$$\langle T^{\mu\nu}\rangle = i\lim_{x'\to x}\Big[-\partial^\mu\partial'^\nu G^{\lambda}{}_{\lambda} + \partial^\mu\partial'_\lambda G^{\lambda\nu} + \partial^\lambda\partial'^\nu G^{\mu}{}_{\lambda} - \partial'^\lambda\partial_\lambda G^{\mu\nu} + \frac{1}{2}\eta^{\mu\nu}\left(\partial^\alpha\partial'_\alpha G^{\lambda}{}_{\lambda} - \partial^\alpha\partial'_\beta G^{\beta}{}_{\alpha}\right)\Big], \tag{22}$$

is the VS of the standard Maxwell SET, and

$$\langle\Xi^{\mu\nu}\rangle = -2ik^{\nu}\epsilon^{\mu\alpha\beta\gamma}\lim_{x'\to x}\partial'_\beta G_{\gamma\alpha}, \tag{23}$$

encodes Lorentz-violating contributions.

5. Casimir effect

Now let us to consider the renormalized VS $\langle\Theta^{\mu\nu}\rangle_{\text{ren}}$, which is obtained as the difference between the VS in the presence of boundaries and that of the vacuum. For two parallel conducting plates separated by a distance D in the z direction, one can construct the renormalized expectation value of the stress-energy tensor, using conservation, tracelessness and symmetry arguments. We simplify our configuration by orienting the coordinate frame so that the plates are perpendicular to the background LV vector $k^{\mu} = (0, \mathbf{k}) = \kappa n^{\mu}$, however the general case proceeds along the same lines.

The explicit calculation of Eqs.(22)-(23), in the case of two parallel plates, allows us to derive $\langle\Theta^{\mu\nu}\rangle_{\text{ren}}$, which is given by

$$\langle\Theta^{\mu\nu}\rangle_{\text{ren}} = \langle t^{\mu\nu}\rangle_{\text{ren}} + (\eta^{\mu\nu} + 4n^{\mu}n^{\nu})\, f_{\text{ren}}(\kappa, z) + n^{\mu}n^{\nu} g_{\text{ren}}(\kappa, z), \tag{24}$$

where $n^{\mu} = (0, 0, 0, 1)$ is the unit normal to the plates,

$$\langle t^{\mu\nu}\rangle_{\text{ren}} = -\frac{\pi^2}{720D^4}\left(\eta^{\mu\nu} + 4n^{\mu}n^{\nu}\right), \tag{25}$$

is the VS in the absence of Lorentz violation [28] and the functions $f(\kappa, z)$ and $g(\kappa, z)$, coming from the calculation of the GF function in Eq. (11), are defined as

$$f(\kappa, z) = \frac{4\kappa^2}{i}\int\frac{d^2\mathbf{p}}{(2\pi)^2}\int\frac{d\omega}{2\pi}\omega^2 \lim_{z'\to z}\left(p^2 + \partial_z\partial'_z\right) M(z, z'), \tag{26}$$

$$g(\kappa, z) = -8i\kappa^2\int\frac{d^2\mathbf{p}}{(2\pi)^2}\int\frac{d\omega}{2\pi}p^2 \lim_{z'\to z} N(z, z'). \tag{27}$$

5.1. Global Casimir energy density

The Casimir energy is defined as the energy per unit area stored in the electromagnetic field between the plates, i.e.

$$\mathcal{E}_C = \int_0^D \langle\Theta^{00}\rangle_{\text{ren}}\, dz, \tag{28}$$

where $\langle\Theta^{00}\rangle_{\text{ren}}$ is the renormalized 00-component of the stress-energy tensor, which according to the procedure described above, it is obtained as the difference between the energy density in the presence of the plates and that of the free vacuum, i.e.

$$\langle\Theta^{00}\rangle_{\text{ren}} = \langle\Theta^{00}\rangle_{\parallel} - \langle\Theta^{00}\rangle_{0}. \tag{29}$$

Here, the labels $\parallel$ and 0 mean that the expectation value is evaluated with the Green's function in the presence of the parallel plates and that of the vacuum, respectively. The final expression for the Casimir energy becomes

$$\mathcal{E}_C = -\frac{\pi^2}{720D^3} + \frac{\kappa^2}{32D}. \tag{30}$$

The Casimir stress is obtained by differentiating the Casimir energy with respect to D, i.e.

$$\mathcal{F}_C = -\frac{d\mathcal{E}_C}{dD} = -\frac{\pi^2}{240D^4} + \frac{\kappa^2}{32D^2}. \tag{31}$$

The first term is recognized as the usual Casimir force in Lorentz-symmetric electrodynamics, while the second represents the Lorentz-violating contribution. We observe that unlike the

$1/D^4$ dependence of the usual Casimir force, the LV part depends on $1/D^2$. As discussed in section 1, the authors in Ref. [18] analyzed the same Casimir configuration within the (3+1)D Chern-Simons extended electrodynamics. They obtained an attractive correction of the form $-\frac{25k_{AF}^2}{720D^2}$, while our result (31) is evidently repulsive. The origin of such disagreement is the Chern-Simons background tensor we work with; while the authors in Ref. [18] used a timelike background $k^\mu = (k_{AF}, \mathbf{0})$, we have chosen the spacelike vector $k^\mu = (0, \mathbf{k})$. Here we point out that the difference between our work and that of Ref. [18] is beyond the background tensor used. For example, they used standard techniques in Casimir physics (zeta function regularization and series summation via the residue theorem), while here we tackle the problem by means of Green's function techniques. The advantage of the local approach over the standard techniques is quite subtle: the former provides information about the local behavior of the quantum fields near a boundary, while the later only allows the calculation of the Casimir force. In 5.2 we discuss the local effects in (3+1)D Chern-Simons electrodynamics, which is the main aim of the present work.

5.2. Local effects

Heretofore, we have considered the global Casimir effect: the total energy of a field configuration or the force per unit area on a bounding surface [29, 30]. Local properties of the quantum vacuum induced by the presence of boundaries are of broad interest in quantum field theory [31] and they must be understood if one is to correctly interpret the inherent divergences in the theory.

The local energy density in Lorentz-symmetric electrodynamics has been discussed extensively in the literature [29, 30, 32]; however, the local effects in Lorentz-violating theories have not been considered. Here we aim to fill in this gap. We begin the analysis by considering an electromagnetic field confined between two parallel conducting plates at $z = 0$ and $z = D$, for which the energy density per unit volume between the plates is

$$\left\langle \Theta^{00} \right\rangle (z) = \left\langle t^{00} \right\rangle (z) + \left\langle \mathbf{T}^{00} \right\rangle (z). \tag{32}$$

A detailed analysis of the local effects due to the first term (in the absence of Lorentz violation) is presented in Ref. [29, 30]. Here we concentrate on the Lorentz-violating contribution,

$$\left\langle \mathbf{T}^{00} \right\rangle (z) = \frac{\kappa^2}{32D^2} \left[1 - 2\csc^2(\pi Z)\right], \qquad Z = \frac{z}{D}. \tag{33}$$

We observe that the z-independent term, $\frac{\kappa^2}{32D^2}$, corresponds to the global renormalized Lorentz-violating energy density obtained in 5.1. The second term encodes the local effects and it can be expressed as follows:

$$\mathcal{S}(Z) \equiv -\frac{\kappa^2}{16D^2}\csc^2(\pi Z). \tag{34}$$

We observe that it diverges quadratically as $z \to 0, D$. Its z integral over the region between the plates diverges linearly. This result reveals a close analogy with the one obtained from the Lorentz-symmetric part. In that case, the singular part diverges quartically as $z \to 0, D$. The less divergent Lorentz-violating contribution (34) can be understood as due to the dimension of Chern-Simons coupling κ. It is worth to mention that integrating this term over z,

$$\int_0^D \mathcal{S}(Z)dz = -\frac{\kappa^2}{16\pi^2}\frac{2}{\Gamma(2)}\int_0^\infty d\rho, \tag{35}$$

we obtain a divergent constant term (D-independent), so it does not contribute to the observable force.

6. Conclusions

Let us summarize the main aims of the present contribution. We analyze the local effects of the quantum vacuum in a particular sector of the minimal SME, namely, the (3+1)D Maxwell-Chern-Simons term with a spacelike Lorentz-violating background field. Concretely, we use Green's function techniques to calculate the Casimir force between two parallel conducting plates separated by a distance D, focusing on the local properties of the quantum vacuum induced by the presence of the boundaries.

One of the contributions of the present work is the construction of the correct (indexed) Green's function for the theory we dealt with. We also analyze the behavior of the local energy density when approaching the plates.

The present work can be further generalized in a variety of ways. For example, the Green's function for different geometries can also be constructed using the same perturbative procedure. On the other hand, our analysis can also be applied to ponderable media. More precisely, we can consider a semi-infinite planar material medium with dielectric constant ε for which the reduced Green's function $\mathfrak{g}_\varepsilon(z,z')$ is known. Now we can use it to evaluate the associated $N_\varepsilon(z,z')$ and $M_\varepsilon(z,z')$ functions, which are what we require to study the Lorentz-violating effects.

Acknowledgments

CAE is supported by the CONACyT Posdoctoral Grant No. 234745.

References

[1] Gambini R and Pullin J 1999 *Phys. Rev.* D **59** 124021
[2] Alfaro J, Morales-Técotl H A and Urrutia L F 2000 *Phys. Rev. Lett.* **84** 2318
[3] Kostelecký V A and Samuel S 1989 *Phys. Rev.* D **39** 683
[4] Kostelecký V A and Potting R 1991 *Nucl. Phys.* B **359** 545
[5] Klinkhamer F R and Rupp C 2004 *Phys. Rev.* D **70** 045020
[6] Martín-Ruiz A 2014 *Phys. Rev.* D **90** 125027
[7] Martín-Ruiz A, Frank A and Urrutia L F 2015 *Phys. Rev.* D **92** 045018
[8] Colladay D and Kostelecký V A 1997 *Phys. Rev.* D **55** 6760
[9] Colladay D and Kostelecký V A 1998 *Phys. Rev.* D **58** 116002
[10] Kostelecký V A and Samuel S 1989 *Phys. Rev.* D **40** 1886
[11] Kostelecký V A and Samuel S 1989 *Phys. Rev.* D **39** 683
[12] Kostelecký V A, Lane C and Pickering A 2002 *Phys. Rev.* D **65** 056006
[13] Kostelecký V A and Lehnert R 2001 *Phys. Rev.* D **63** 065008
[14] Carroll S M, Field G B and Jackiw R 1990 *Phys. Rev.* D **41** 1231
[15] Casimir H B G 1948 *Proc. K. Ned. Akad. Wet.* **51** 793
[16] Bressi G, Carugno G, Onofrio R and Ruoso G 2002 *Phys. Rev. Lett.* **88** 041804
[17] Frank M and Turan I 2006 *Phys. Rev.* D **74** 033016
[18] Kharlanov O G and Zhukovsky V 2010 *Phys. Rev.* D **81** 025015
[19] Martín-Ruiz A and Escobar C A 2016 *Phys. Rev.* D **94** 076010
[20] Martín-Ruiz A and Escobar C A 2017 *Phys. Rev.* D **95** 036011
[21] Adam C and Klinkhamer F R 2001 *Nuc. Phys.* B **607** 247
[22] Alfaro J, Cambiaso M, Giacconi P and Soldati R 2010 *Int. J. Mod. Phys.* A **25** 3271
[23] Colladay D, McDonald P, Noordmans J and Potting R 2017 *Phys. Rev.* D **95** 025025
[24] Casana R, Ferreira Jr M and Santos C E H 2008 *Phys. Rev.* D **78** 025030
[25] Belich Jr H, Ferreira Jr M, Helayël-Neto J A and Orlando M T D 2003 *Phys. Rev.* D **68** 025005
[26] Lehnert R and Potting R 2004 *Phys. Rev. Lett.* **93** 110402
[27] Kaufhold C and Klinkhamer F R 2007 *Phys. Rev.* D **76** 025024
[28] Brown L S and Maclay G J 1969 *Phys. Rev.* **184** 1272
[29] Milton K A 2001 *The Casimir effect: Physical Manifestation of Zero-Point Energy* (Singapore: World Scientific)
[30] Bordag M, Klimchitskaya G L, Mohideen U and Mostepanenko V M 2009 *Advances in the Casimir effect* (Oxford: Oxford University Press)
[31] Deutsch D and Candelas P 1979 *Phys. Rev.* D **20** 3063
[32] Sopova V and Ford L H 2002 *Phys. Rev.* D **66** 045026

IOP Conf. Series: Journal of Physics: Conf. Series **952** (2018) 012011 doi:10.1088/1742-6596/952/1/012011

Quantization of Space-like States in Lorentz-Violating Theories

Don Colladay
New College of Florida, Sarasota, FL 34234, USA
E-mail: colladay@ncf.edu

Abstract. Lorentz violation frequently induces modified dispersion relations that can yield space-like states that impede the standard quantization procedures. In certain cases, an extended Hamiltonian formalism can be used to define observer-covariant normalization factors for field expansions and phase space integrals. These factors extend the theory to include non-concordant frames in which there are negative-energy states. This formalism provides a rigorous way to quantize certain theories containing space-like states and allows for the consistent computation of Cherenkov radiation rates in arbitrary frames and avoids singular expressions.

1. Introduction

Theories involving space-like solutions to the dispersion relation can yield negative-energy states in certain reference frames. The quantization of such a theory in these frames is known to be problematic. For example, the simple tachyonic neutrino model [1], proposed to accommodate anomalous neutrino mass measurement results, defined by the Lagrangian

$$L = i\overline{\psi}\gamma_5 \partial\!\!\!/ \psi - m_\nu \overline{\psi}\psi, \tag{1}$$

leads to the dispersion relation $p^2 = -m_\nu^2$. Conventional attempts at quantization typically fail due to the impossibility of finding a reference frame-independent separation of particle and anti-particle states. This leads to inconsistencies in the usual re-interpretation of the negative-energy states in terms of anti-particles. The situation is demonstrated in figures 1 and 2 which shows the dispersion relation plotted in two different reference frames. Note that some of the anti-particle states in one frame are transformed into particle states in the other frame.

Similar issues can occur in Lorentz-violating theories, such as in the CPT-violating, massive photon model (with k_{AF}^μ a constant background vector field)

$$\mathcal{L}_A = -\tfrac{1}{4}F_{\mu\nu}F^{\mu\nu} + \tfrac{1}{2}k_{AF}^\kappa \epsilon_{\kappa\lambda\mu\nu}A^\lambda F^{\mu\nu} + \tfrac{1}{2}m_\gamma^2 A_\mu A^\mu - \frac{1}{2\xi}(\partial_\mu A^\mu)^2, \tag{2}$$

with momentum-space solution containing the perturbed dispersion relation

$$R_T(p) = \frac{1}{4}(p^2 - m_\gamma^2)^2 - (p \cdot k_{AF})^2 + p^2 k_{AF}^2 = 0, \tag{3}$$

which has the observer-covariant factorization $R_T(p) = R_+(p)R_-(p)$ with

$$R_\pm(p) = \frac{1}{2}(p^2 - m_\gamma^2) \pm \sqrt{(p \cdot k_{AF})^2 - p^2 k_{AF}^2}. \tag{4}$$

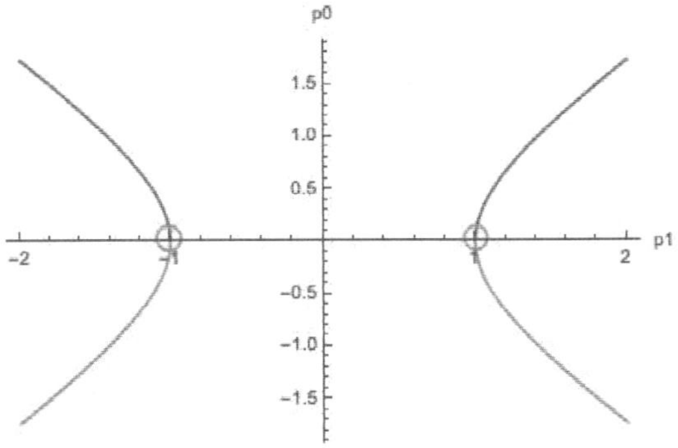

Figure 1. Plot of dispersion relation with separation between particle and anti-particle states.

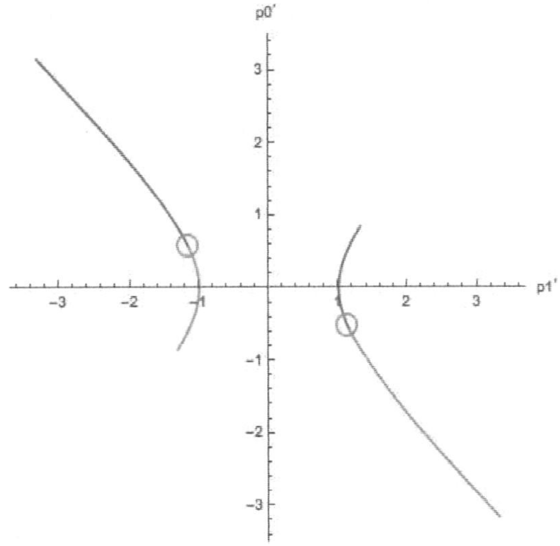

Figure 2. Plot of dispersion relation in boosted frame where some of particle states have $p_0' < 0$.

This factorization can be directly related to solutions to the modified Dirac equation "off-shell" spinor solutions

$$\frac{1}{2}(\not{p} - m - b_\mu \gamma_5 \gamma^\mu) u_\pm(p) = \mathcal{R}_\pm(p) u_\pm(p), \tag{5}$$

with $\mathcal{R}_\pm(p) = \frac{1}{2}(p^2 - m^2 - b^2) \pm \sqrt{(b \cdot p)^2 - b^2 p^2}$ as above acting on spinors $u_\pm = (\not{p} + m - \gamma_5 \not{b}) w_\pm$ giving the condition

$$\frac{1}{2} \epsilon_{\mu\nu\alpha\beta} \sigma^{\mu\nu} p^\alpha b^\beta w_\pm = \pm\sqrt{(b \cdot p)^2 - b^2 p^2} w_\pm, \tag{6}$$

which demonstrates that they are eigenstates of the Pauli-Lubanski vector.

The factor $R_+(p) = 0$ has space-like solutions with $p^2 < 0$ at high-momenta. Figure 3 shows

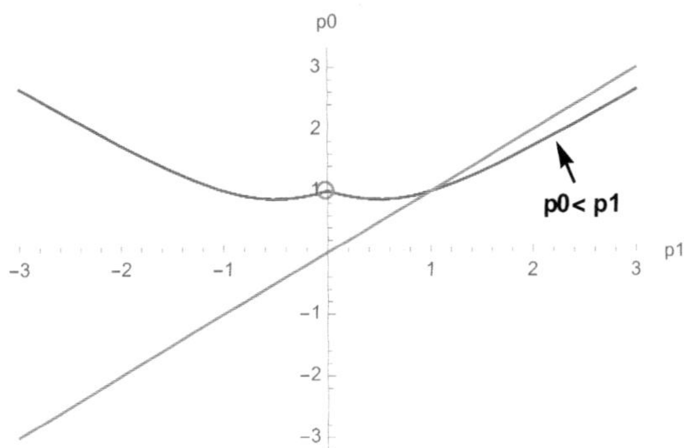

Figure 3. Space-like states for dispersion relation $R_+(p) = 0$ compared to conventional massless photon.

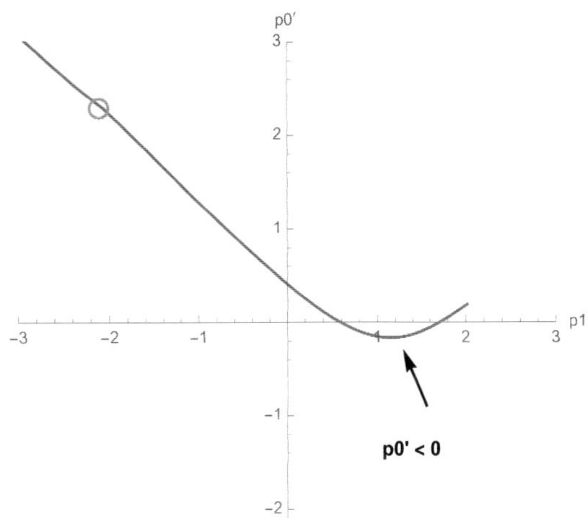

Figure 4. Same states in a highly boosted frame where some energies are negative.

the positive-energy particle states for the case of a pure timelike b^μ coefficient. Space-like states exist in the region where the dispersion relation plot dips below the light-cone. Figure 4 displays the energy and momentum in a highly boosted frame where the energy dips below zero in some region. A first attempt may be to re-interpret these negative-energy solutions as anti-particles and use conventional field normalization factors, but it turns out that this procedure leads to singular factors in the commutation relations that makes the quantization procedure suspect. Instead, an alternative procedure which makes use of an extended Hamiltonian formalism will be used that makes all of the states into particle states with an observer covariant interpretation of particle states that is valid in all frames. To identify the appropriate normalization factors, use is made of the classical mechanical limit of the theory.

2. Classical Mechanics Lagrangians in the SME

The traditional method of computing Lagrangians in the SME [2] uses a Legendre transformation in conjunction with the group velocity to obtain the appropriate Lagrange functions from the dispersion relation. The process can be made to appear covariant by introducing an arbitrary path parametrization and a four-velocity $u^\mu = dx^\mu/d\lambda$. For example, the fermion dispersion relation with an external b^μ-parameter considered in Ref. [2] leads to the dispersion relation of the CPT-violating massive photon model (see Eqs. (2) and (3)) by making the replacements $b^\mu \to k^\mu_{AF}$, and $m^2 \to m_\gamma^2 - k^2_{AF}$, which preserves hermiticity provided $m_\gamma^2 > k^2_{AF}$. The relevant dispersion relation is

$$\mathcal{R}(p) = \frac{1}{4}(p^2 - m^2 + b^2)^2 - (b \cdot p)^2 + m^2 b^2 = 0. \tag{7}$$

The group velocity $\vec{v} = d\vec{x}/dt$ is computed using implicit differentiation as

$$v_i = \frac{\partial p_0}{\partial p^i}, \tag{8}$$

which is well-defined away from any singular points where the energy surfaces become degenerate. The relevant Lagrangians are computed using the Legendre transformation $L = \vec{u} \cdot \vec{p} - p^0(\vec{p})$. The expression for $\vec{v}(\vec{p})$ defined in Eq. (8) is inverted for $\vec{p}(\vec{v})$ to give

$$L_\pm[\vec{v}, x] = -m\sqrt{1 - \vec{v}^2} \mp \sqrt{(b^0 - \vec{b} \cdot \vec{v})^2 - b^2(1 - \vec{v}^2)}, \tag{9}$$

where the signs represent two valid solutions that reduce to the standard case when $b^\mu \to 0$.

The above procedure can be made explicitly covariant by introducing a zero-component of the four-velocity, $u^0(\lambda)$, and an arbitrary parametrization $\lambda(t)$ so that the action appears covariant in terms of the four-velocity $u^\mu = dx^\mu/d\lambda$, as

$$L_\pm[u^\mu, x] = -m\sqrt{u^2} \mp \sqrt{(b \cdot u)^2 - b^2 u^2}. \tag{10}$$

The above approach has some issues which make a relativistic Hamiltonian description difficult. For one, the function u^μ has a "gauge" degree of freedom due to the re-parametrization invariance and therefore it is not fully determined by the equation of motion. Computing the momentum $p^\mu = -\frac{\partial L}{\partial u_\mu}$ gives

$$p^\mu = \frac{m u^\mu}{\sqrt{u^2}} \pm \frac{(u \cdot b) b^\mu - b^2 u^\mu}{\sqrt{(b \cdot u)^2 - b^2 u^2}}, \tag{11}$$

which is insensitive to a scaling of the four-velocity. This means that Eq. (11) is not invertible to find $u(p)$ unless some condition on u^μ is enforced (such as $u^2 = 1$ when $\lambda = \tau$ is taken as the proper time). Calculation of the relativistic Hamiltonian gives

$$\mathcal{H} = p^0 u^0 - \vec{p} \cdot \vec{u} - L = 0, \tag{12}$$

IOP Conf. Series: Journal of Physics: Conf. Series **952** (2018) 012011 doi:10.1088/1742-6596/952/1/012011

since $L = -u_\mu p^\mu$, yielding no useful Hamiltonian formalism. It is also unclear how the Legendre transformation specifically maps the Lagrange functions L_+ and L_- to the hypersurfaces defined by $R_+(p) = 0$ and $R_-(p) = 0$.

3. Extended Hamiltonian Formalism

The relativistic Hamiltonian can be extended to "off-shell" values using an extended Hamiltonian formalism originally due to Dirac. The procedure is to introduce a new variable $e(\lambda)$ as a Lagrange multiplier [3]

$$S_\pm^* = -\int \left[me^{-1}u^2 \pm \sqrt{(b\cdot u)^2 - b^2u^2} - \frac{e}{m}\mathcal{R}_\mp(p,x)\right] d\lambda, \tag{13}$$

with the observer covariant factorization

$$\mathcal{R}(p) = \mathcal{R}_+(p)\mathcal{R}_-(p), \tag{14}$$

and

$$\mathcal{R}_\pm = \frac{1}{2}\left(p^2 - m^2 - b^2\right) \pm \sqrt{(b\cdot p)^2 - b^2p^2}, \tag{15}$$

which agrees with the previous action "on-shell" where $\mathcal{R}_\pm = 0$, but modified action applies for unconstrained variations of u^μ and p^μ yielding an extended relativistic Hamiltonian

$$\mathcal{H}_\pm^* = -\frac{e}{m}\mathcal{R}_\mp(p,x) = -\frac{e}{2m}\left(p^2 - m^2 - b^2 \pm 2\sqrt{(b\cdot p)^2 - b^2p^2}\right), \tag{16}$$

and one of Hamilton's equations gives the velocity as

$$u^\mu = -\frac{\partial \mathcal{H}^*}{\partial p_\mu} = \frac{e}{m}\left(p^\mu \mp \frac{(b\cdot p)b^\mu - b^2p^\mu}{\sqrt{(b\cdot p)^2 - b^2p^2}}\right), \tag{17}$$

which can be inverted to give

$$p^\mu = \frac{mu^\mu}{e} \pm \frac{(u\cdot b)b^\mu - b^2u^\mu}{\sqrt{(b\cdot u)^2 - b^2u^2}}, \tag{18}$$

providing a well-defined Legendre transformation between the Lagrangian and extended Hamiltonian functions, at least away from singular points. Behavior near singular points is discussed further in Refs. [3, 4].

4. Quantization

A typical field expansion contains the observer-covariant phase space factor

$$\int \frac{d^3\vec{p}}{2p^0(\vec{p})} = \int d^4p\ \delta(p^2 - m^2)\Theta(p^0), \tag{19}$$

which is problematic in frames where space-like states satisfy $p^0 \le 0$ and the naive commutators

$$[a(p), a^\dagger(p)] = (2\pi)^3 2p^0(\vec{p})\delta(\vec{p} - \vec{p}'), \tag{20}$$

vanish or go negative, leading to interpretational difficulties. To circumvent this problem, the correct relativistic Hamiltonian for the theory in Eq. (16) can be used in the delta function to put the particles properly on-shell according to their classical mechanical limit. In this way the phase-space integral becomes

$$\int d^4p \left(\frac{-e}{2m}\right) \delta(\mathcal{H}^*_\pm(p)) = \int \frac{d^3\vec{p}}{\Lambda^{0\prime}_\pm(p)}, \tag{21}$$

with

$$\Lambda^{0\prime}_\pm(p) = -\left(\frac{2m}{e}\right) \frac{\partial \mathcal{H}^*_\pm}{\partial p^0} = 2\left(p^0 \pm \frac{(b\cdot p)b^0 - b^2 p^0}{\sqrt{(b\cdot p)^2 - b^2 p^2}}\right), \tag{22}$$

which enforces the correct dynamics on surface $\mathcal{H}^*_\pm = 0$. Since $u^2 = u_0^2 - \vec{u}^2 = 1$ (in proper time parameterization), the factor $u^0 = -\partial \mathcal{H}^*_\pm / \partial p^0 > 0$ in all frames, so that the denominator in the phase-space factor in Eq. (21) is positive definite. Using this factor in the field expansions (with the appropriate photon extended Hamiltonian [5]) yields

$$A^\mu(x) = \int \sum_\pm \frac{d^3\vec{p}}{\Lambda^{0\prime}_\pm(p)} \epsilon^\mu a^\dagger(p) e^{-ip\cdot x} + c.c., \tag{23}$$

and the corresponding phase space factors are well-defined in all frames, not just concordant [6] ones in which the Lorentz-violating parameters are small. Defining the canonical momenta $\pi^\mu = \partial\mathcal{L}/\partial\dot{A}_\mu$ gives

$$\pi^\mu(x) = F^{\mu 0}(x) + \epsilon^{0\mu\alpha\beta}(k_{AF})_\alpha A_\beta(x) - \eta^{\mu 0}\tfrac{1}{\xi}\partial_\nu A^\nu(x). \tag{24}$$

Imposing the canonical quantization rules

$$[A_\mu(t,\vec{x}), \pi^\nu(t,\vec{y})] = i\delta^\nu_\mu \delta^3(\vec{x}-\vec{y}), \quad [A_\mu(t,\vec{x}), A_\nu(t,\vec{y})] = 0, \tag{25}$$

yields the momentum-space state algebra

$$[a(\vec{p}), a^\dagger(\vec{p}^{\,\prime})] = (2\pi)^3 \Lambda^{0\prime}_\pm(p) \delta(\vec{p} - \vec{p}^{\,\prime}), \tag{26}$$

which has a positive-definite $\Lambda^{0\prime}_\pm(p) > 0$ in all observer frames.

5. Summary

Lorentz-violating field theories that lead to space-like states are not automatically ruled out at the quantum level, even though there may exist frames in which the energy runs negative. By using an appropriate modified Hamiltonian formalism, it is possible (at least in certain specific cases) to define positive-definite phase space factors that can be used to expand the fields and impose a consistent separation between particle and anti-particle states. The corresponding phase-space factors are important in consistent calculations like Cherenkov processes [7, 8].

6. Acknowledgments

I would like to thank the New College faculty development summer funds program for funding.

References

[1] Chodos A, Hauser A I and Kostelecky V A 1985 *Phys. Lett.* B **150** 431
[2] Kostelecky V A and Russell N 2010 *Phys. Lett.* B **693** 443
[3] Colladay D 2017 *Phys. Lett.* B **772** 694
[4] Colladay D and McDonald P 2015 *Phys. Rev.* D **92** 085031
[5] Colladay D, McDonald P, Noordmans J P and Potting R 2017 *Phys. Rev.* D **95** 025025
[6] Lehnert R and Kostelecky V A 2000 *Phys. Rev.* D **63** 065008
[7] Colladay D, McDonald P and Potting R 2016 *Phys. Rev.* D **93** 125007
[8] Colladay D, Noordmans J P and Potting R 2017 *Phys. Rev.* D **96** 035034

IOP Conf. Series: Journal of Physics: Conf. Series **952** (2018) 012012 doi:10.1088/1742-6596/952/1/012012

Consistent description of field theories with non-Hermitian mass terms

Jean Alexandre[1], **Peter Millington**[2] **and Dries Seynaeve**[1]

[1] Department of Physics, King's College London, London WC2R 2LS, UK

[2] School of Physics and Astronomy, University of Nottingham, Nottingham NG7 2RD, UK

Abstract. We review how to describe a field theory that includes a non-Hermitian mass term in the region of parameter space where the Lagrangian is $\mathcal{PT}$-symmetric. The discrete symmetries of the system are essential for understanding the consistency of the model, and the link between conserved current and variation of the Lagrangian has to be revisited in the case of continuous symmetries.

1. Introduction

Among the potential extensions of the Standard Model is the possibility to consider non-Hermitian operators, as long as the corresponding energies remain real and the evolution remains unitary. It is known in Quantum Mechanics that a Hamiltonian that is symmetric under the combined action of parity ($\mathcal{P}$) and time reversal ($\mathcal{T}$) has real eigenvalues [1], and we review here some recent work done in scalar and fermionic field theories that involve non-Hermitian mass terms. In the region of parameter space where the Lagrangian is $\mathcal{PT}$-symmetric, we show that a consistent description of such a system can be achieved if one associates the degrees of freedom with the $\mathcal{PT}$-conjugate fields, rather than the Hermitian-conjugate ones. After reviewing two different non-interacting models, we explain some features of a non-Hermitian extension of Quantum Electrodynamics.

2. Anti-Hermitian mass terms

The scalar and fermion models that we consider are described by the Lagrangians

$$\begin{aligned}\mathcal{L}_s &= \begin{pmatrix}\partial_\nu\phi_1^\star & \partial_\nu\phi_2^\star\end{pmatrix}\begin{pmatrix}\partial^\nu\phi_1\\ \partial^\nu\phi_2\end{pmatrix} - \begin{pmatrix}\phi_1^\star & \phi_2^\star\end{pmatrix}\begin{pmatrix}m_1^2 & \mu^2\\ -\mu^2 & m_2^2\end{pmatrix}\begin{pmatrix}\phi_1\\ \phi_2\end{pmatrix} \quad \text{and}\\ \mathcal{L}_f &= \overline{\psi}\left(i\partial\!\!\!/ - m - \mu\gamma^5\right)\psi\,, \end{aligned} \tag{1}$$

and contain the anti-Hermitian terms $\mu^2(\phi_1^\star\phi_2 - \phi_1\phi_2^\star)$ and $\mu\overline{\psi}\gamma^5\psi$, respectively.

In the scalar case [2], the eigenvalues of the mass matrix are

$$M_{s\pm}^2 = \frac{1}{2}(m_1^2 + m_2^2) \pm \frac{1}{2}\sqrt{(m_1^2 - m_2^2)^2 - 4\mu^4}\,, \tag{2}$$

and these lead to real energies as long as $2\mu^2 \leq |m_1^2 - m_2^2|$. The corresponding eigenvectors are not orthogonal with respect to Hermitian conjugation and become parallel in the limit where

IOP Conf. Series: Journal of Physics: Conf. Series **952** (2018) 012012 doi:10.1088/1742-6596/952/1/012012

the eigenmasses become degenerate. At this degenerate point, the system describes half of the original degrees of freedom.

In the fermionic case, originally introduced in [3] and also related to [4], the energies are real as long as $|\mu| \leq m$, and the mass is

$$M_f = \sqrt{m^2 - \mu^2} \ . \tag{3}$$

It is interesting to see that, in the limits $\mu \to \pm m$ where the fermion is effectively massless, one loses either the right-chiral component or the left-chiral component. Indeed, in terms of

$$\psi_R = \frac{1}{2}(1+\gamma^5)\psi \quad \text{and} \quad \psi_L = \frac{1}{2}(1-\gamma^5)\psi \ , \tag{4}$$

the Lagrangian reads

$$\mathcal{L}_f = \overline{\psi}_R i\partial\!\!\!/ \psi_R + \overline{\psi}_L i\partial\!\!\!/ \psi_L - \overline{\psi}_L(m+\mu)\psi_R - \overline{\psi}_R(m-\mu)\psi_L \ . \tag{5}$$

The limit $\mu = m$, for example, leads to the equation of motion $i\partial\!\!\!/ \psi_L = 2m\psi_R$ and therefore $\mathcal{L}_f^{\text{on-shell}} = \overline{\psi}_R i\partial\!\!\!/ \psi_R$. Similarly, the limit $\mu = -m$ leads to $\mathcal{L}_f^{\text{on-shell}} = \overline{\psi}_L i\partial\!\!\!/ \psi_L$.

In both the scalar and fermionic cases, one then reaches a singularity in some specific limits for μ, since one loses half of the degrees of freedom. These "exceptional points" correspond to the boundaries, in parameter space, beyond which the $\mathcal{PT}$ symmetry is broken.

3. $\mathcal{PT}$ symmetry and equations of motion

A generic feature of non-Hermitian models with $\mathcal{PT}$ symmetry is their interpretation in terms of coupled systems with gain and loss. For the scalar model, one can see from the conserved current (see the next section) that one field plays the role of source and the other plays the role of sink; $\mathcal{P}$ swaps the fields, whereas $\mathcal{T}$ swaps gain and loss. For this picture to be consistent, one field must transform as a pseudo-scalar [2], and the $\mathcal{PT}$-conjugate transpose of the scalar doublet is

$$\Phi^\ddagger = \left[\begin{pmatrix} 1 & 0 \\ 0 & -1 \end{pmatrix} \Phi^\star\right]^\mathsf{T} = (\phi_1^\star, -\phi_2^\star) \quad \text{where} \quad \Phi = \begin{pmatrix} \phi_1 \\ \phi_2 \end{pmatrix} \ . \tag{6}$$

The scalar Lagrangian can then be written in an explicitly $\mathcal{PT}$-symmetric way as

$$\mathcal{L}_s = \Phi^\ddagger \begin{pmatrix} -\Box - m_1^2 & -\mu^2 \\ -\mu^2 & \Box + m_2^2 \end{pmatrix} \Phi \ , \tag{7}$$

such that the equations of motion are obtained from the variational principle

$$\frac{\delta S_s}{\delta \Phi^\ddagger} = 0 \quad \text{or} \quad \left(\frac{\delta S_s}{\delta \Phi}\right)^\ddagger = 0 \ , \tag{8}$$

where S_s is the $\mathcal{PT}$-symmetric action.

The $\mathcal{PT}$-conjugate transpose of the fermion field is

$$\psi^\ddagger = [i\gamma^0\gamma^1\gamma^3\psi^\star]^\mathsf{T} = i\overline{\psi}\gamma^3\gamma^1 \ , \tag{9}$$

and the explicitly $\mathcal{PT}$-symmetric form of the fermionic Lagrangian is

$$\mathcal{L}_f = -i\psi^\ddagger(x)\gamma^1\gamma^3(i\overleftrightarrow{\partial\!\!\!/} - m - \mu\gamma^5)\psi(x) \ , \tag{10}$$

IOP Conf. Series: Journal of Physics: Conf. Series **952** (2018) 012012 doi:10.1088/1742-6596/952/1/012012

where

$$\overleftrightarrow{\not\partial} \equiv \frac{1}{2}\left(\overrightarrow{\not\partial} - \overleftarrow{\not\partial}\right) . \tag{11}$$

The equations of motion are then obtained from the variational principle

$$\frac{\delta S_f}{\delta \psi^\ddagger} = 0 \quad \text{or} \quad \left(\frac{\delta S_f}{\delta \psi}\right)^\ddagger = 0 . \tag{12}$$

Equivalent equations of motion for the scalar and the fermion models are obtained from the functional variations

$$\begin{array}{lccc} \text{(scalar)} & \dfrac{\delta S_s}{\delta \Phi^\star} = 0 & \text{or} & \dfrac{\delta S_s^\star}{\delta \Phi} = 0 \\ \text{(fermion)} & \dfrac{\delta S_f}{\delta \overline{\psi}} = 0 & \text{or} & \dfrac{\delta S_f^\star}{\delta \psi} = 0 . \end{array} \tag{13}$$

In addition, we note that one could instead choose the equations of motion to be defined via

$$\begin{array}{lccc} \text{(scalar)} & \dfrac{\delta S_s^\star}{\delta \Phi^\star} = 0 & \text{or} & \dfrac{\delta S_s}{\delta \Phi} = 0 \\ \text{(fermion)} & \dfrac{\delta S_f^\star}{\delta \overline{\psi}} = 0 & \text{or} & \dfrac{\delta S_f}{\delta \psi} = 0 , \end{array} \tag{14}$$

which would correspond to the change $\mu^2 \to -\mu^2$ (scalar case) or $\mu \to -\mu$ (fermion case). This would, however, not change the physical predictions, since it is equivalent to interchanging $\phi_1 \leftrightarrow \phi_2$ for the scalar case and $\psi_R \leftrightarrow \psi_L$ for the fermionic case.

4. Conserved currents

By making use of the equations of motion, one can show that the conserved current for the scalar model is

$$j_s^\nu = i\left(\phi_1^\star \partial^\nu \phi_1 - \phi_1 \partial^\nu \phi_1^\star\right) - i\left(\phi_2^\star \partial^\nu \phi_2 - \phi_2 \partial^\nu \phi_2^\star\right) , \tag{15}$$

which corresponds to the phase transformation [2]

$$\Phi' = \exp\left[+i\alpha\begin{pmatrix}1 & 0\\ 0 & -1\end{pmatrix}\right]\Phi \quad ; \quad \Phi^{\ddagger\prime} = \Phi^\ddagger \exp\left[-i\alpha\begin{pmatrix}1 & 0\\ 0 & -1\end{pmatrix}\right] . \tag{16}$$

Notice that the sign difference between the components of the doublet reflects the source/sink behaviour. For the fermionic case, the conserved current is [6]

$$j_f^\nu = \overline{\psi}\gamma^\nu\left(1 + \frac{\mu}{m}\gamma^5\right)\psi , \tag{17}$$

which corresponds to the phase transformation [2]

$$\psi' = \exp\left[+i\alpha\left(1 + \frac{\mu}{m}\gamma^5\right)\right]\psi \quad ; \quad \overline{\psi}' = \overline{\psi}\exp\left[-i\alpha\left(1 - \frac{\mu}{m}\gamma^5\right)\right] . \tag{18}$$

However, in both cases, the phase transformations (16) or (18) do not leave the respective Lagrangian invariant, and the usual link between invariance of the Lagrangian and current conservation does not hold [2].

The reason for this becomes clear when one considers the full structure of the variation of the Lagrangian under a field transformation, which in the scalar case reads

$$\delta\mathcal{L}_s = \left(\frac{\partial\mathcal{L}_s}{\partial\Phi} - \partial_\nu\frac{\partial\mathcal{L}_s}{\partial(\partial_\nu\Phi)}\right)\delta\Phi + \delta\Phi^\ddagger\left(\frac{\partial\mathcal{L}_s}{\partial\Phi^\ddagger} - \partial_\nu\frac{\partial\mathcal{L}_s}{\partial(\partial_\nu\Phi^\ddagger)}\right) + \partial_\nu(\delta j^\nu) \ . \tag{19}$$

Since the theory is not Hermitian, the two parentheses on the right-hand side of the previous equation cannot simultaneously vanish on-shell. If we choose the equations of motion to be defined by eq.(8) then current conservation requires

$$\delta\mathcal{L}_s = \left(\frac{\partial\mathcal{L}_s}{\partial\Phi} - \partial_\nu\frac{\partial\mathcal{L}_s}{\partial(\partial_\nu\Phi)}\right)\delta\Phi = 2\mu^2(\phi_2^\star\delta\phi_1 - \phi_1^\star\delta\phi_2) \ . \tag{20}$$

Similarly, current conservation in the fermionic case requires that the Lagrangian variation is

$$\delta\mathcal{L}_f = \left(\frac{\partial\mathcal{L}_f}{\partial\psi} - \partial_\nu\frac{\partial\mathcal{L}_f}{\partial(\partial_\nu\psi)}\right)\delta\psi = -2\mu\overline{\psi}\gamma^5\delta\psi \ . \tag{21}$$

One cannot therefore have a conserved current together with invariance of the non-Hermitian part of the Lagrangian. Current conservation nevertheless remains the essential physical feature, such that invariance of the Lagrangian is not required, as it would be for an Hermitian theory. Instead, the required transformation of the Lagrangian is fixed by an identity of the form (20) or (21), and, in the next section, we give an alternative derivation for the fermionic conserved current, which does not rely on the variational procedure [2].

5. Non-unitary map

The fermionic equation of motion can be written in the Schrödinger form

$$i\partial_0\psi = \gamma_0(\vec{\gamma}\cdot\vec{p} + m + \mu\gamma^5)\psi \ , \tag{22}$$

and one can look for a map $\chi = B\psi$ that leads to

$$i\partial_0\chi = \gamma_0(\vec{\gamma}\cdot\vec{p} + M_f)\chi \quad \text{with } M_f = \sqrt{m^2-\mu^2} \ . \tag{23}$$

The latter equation of motion arises from the Hermitian Lagrangian

$$\mathcal{L}_\chi = \overline{\chi}(i\partial\!\!\!/ - M_f)\chi \ , \tag{24}$$

for which we know that the $U(1)$ conserved current is $j_f^\nu = \overline{\chi}\gamma^\nu\chi$. Expressed in terms of the original field, the current is then

$$j_f^\nu = \overline{\psi}\gamma^0 B^\dagger\gamma^0\gamma^\nu B\psi \ . \tag{25}$$

The transformation matrix B is not unitary, since it maps an equation of motion obtained from a non-Hermitian Lagrangian to an equation of motion obtained from a Hermitian Lagrangian. We can fix the form of B, since it must satisfy

$$B\gamma^0(\vec{\gamma}\cdot\vec{p} + m + \mu\gamma^5)B^{-1} = \gamma^0(\vec{\gamma}\cdot\vec{p} + M_f) \ , \tag{26}$$

for any momentum $\vec{p}$, and it is easy to find that

$$B \propto 1 + \gamma^5\sqrt{\frac{1-\sqrt{1-\mu^2/m^2}}{1+\sqrt{1-\mu^2/m^2}}} \ . \tag{27}$$

Together with eq. (25), we recover the conserved current (17). Note that the singularity of the limit $\mu^2 \to m^2$ can be seen here because B becomes proportional to the projector $1+\gamma^5$ and thus has no inverse.

6. Non-Hermitian extension of QED

The gauged model for the fermionic Lagrangian is [5]

$$\mathcal{L} = -\frac{1}{4}F^{\rho\sigma}F_{\rho\sigma} + \overline{\psi}\left[i\partial\!\!\!/ - A\!\!\!/(g_V + g_A\gamma^5) - m - \mu\gamma^5\right]\psi\ , \tag{28}$$

where $F_{\rho\sigma} = \partial_\rho A_\sigma - \partial_\sigma A_\rho$. In the massless case ($m = \mu = 0$), the action is invariant under the combined vector plus axial-vector gauge transformations

$$\begin{aligned} A_\rho &\longrightarrow A'_\rho = A_\rho - \partial_\rho\theta \\ \psi &\longrightarrow \psi' = \exp\left[i\left(g_V + g_A\gamma^5\right)\theta\right]\psi \\ \overline{\psi} &\longrightarrow \overline{\psi}' = \overline{\psi}\exp\left[i\left(-g_V + g_A\gamma^5\right)\theta\right]\ . \end{aligned} \tag{29}$$

For $0 < |\mu| \leq m$, the one-loop corrections in dimension $4 - 2\epsilon$ can be calculated in the chiral basis, and the divergent part of the polarisation tensor is [5]

$$\Pi^{\rho\sigma} = \frac{g_V^2 + g_A^2}{12\pi^2\epsilon}(p^\rho p^\sigma - p^2\eta^{\rho\sigma}) + \frac{g_A^2}{\pi^2\epsilon}(m^2 - \mu^2)\eta^{\rho\sigma}\ , \tag{30}$$

becoming transverse when $\mu^2 \to m^2$. In this limit, the full vector plus axial-vector gauge symmetry is recovered, and this is consistent with our earlier observation (in section 2) that the theory effectively becomes massless.

This model may have interesting implications for neutrino physics [5], since the probability density that one obtains from the current (17) is

$$j^0 = \left(1 + \frac{\mu}{m}\right)|\psi_R|^2 + \left(1 - \frac{\mu}{m}\right)|\psi_L|^2\ . \tag{31}$$

Thus, the contribution from the right-handed component, as well as the fermion mass, can be made arbitrarily small by choosing the ratio $\mu/m \simeq -1$. Note that this result is consistent with the study [7] of a non-Hermitian system of fermions on the lattice, where the numbers of right- and left-handed fermions are not the same. A more detailed description of a non-Hermitian gauge-Yukawa model is given in [5] (see also [8]).

7. Conclusion

As shown here, the description of a non-Hermitian field theory that is $\mathcal{PT}$-symmetric is consistent if one considers the $\mathcal{PT}$-conjugate fields to be the relevant degrees of freedom, instead of the Hermitian-conjugate fields. Further studies of interacting theories, involving non-perturbative tools, are now being considered, specifically the Schwinger-Dyson or the Wilsonian approaches. For this, the consistency of the path-integral quantisation necessarily requires us to integrate over $\mathcal{PT}$-conjugate pairs of degrees of freedom, consistent with the variational procedures described above.

Acknowledgements The work of PM is supported by STFC Grant No. ST/L000393/1 and a Leverhulme Trust Research Leadership Award.

References

[1] Bender C M 2005 *Contemp. Phys.* **46** 277
[2] Alexandre J, Millington P and Seynaeve D 2017 *Phys. Rev.* D **96** 065027
[3] Bender C M, Jones H F and Rivers R J 2005 *Phys. Lett.* B **625** 333
[4] Jone-Smith K and Matur H 2010 *Phys. Rev.* A **82** 042101
[5] Alexandre J, Bender C M and Millington P 2015 *J. High Energy Phys.* **1511** 111
[6] Alexandre J and Bender C M 2015 *J. Phys.* A **48** 185403
[7] Chernodub M N 2017 *J. Phys.* A **50** 385001
[8] Alexandre J, Bender C M and Millington P 2017 *J. Phys.: Conf. Series* **873** 012047

IOP Conf. Series: Journal of Physics: Conf. Series **952** (2018) 012013 doi:10.1088/1742-6596/952/1/012013

Vector models with spontaneous Lorentz-symmetry breaking

C A Escobar[1] **and L F Urrutia**[2]

[1] CENTRA, Departamento de Física, Universidade do Algarve, 8005-139 Faro, Portugal
[2] Instituto de Ciencias Nucleares, Universidad Nacional Autónoma de México, 04510 CDMEX, México

E-mail: `cruiz@ualg.pt`, `urrutia@nucleares.unam.mx`

Abstract. Even though models with spontaneous Lorentz-symmetry breaking also damage gauge invariance, an interesting possibility that emerges is to interpret the resultant massless Goldstone bosons as the gauge bosons of the related gauge theory. In this contribution we review the conditions under which gauge invariance is recovered from such models. To illustrate our general approach we consider the classical Abelian bumblebee and Nambu models. In the former case we prove its connection with electrodynamics by a procedure which takes proper care of the gauge-fixing conditions. In the case of the Abelian Nambu model its relation with electrodynamics is established in such a way that the generalization to the non-Abelian case is straightforward.

1. Introduction

An alternative way of looking at global active Lorentz-symmetry breaking (LSB) is to consider its spontaneous version, which yields Goldstone bosons (GBs) of tensorial character as opposed to the standard Higgs mechanism with scalar content. These GBs can be viewed as realizations of photons, gluons and gravitons, for example. One of the main questions posed by this approach is to determine the sector in the phase space where the model with broken Lorentz symmetry reduces to the gauge theory, to be called the mother gauge theory (MGT), the gauge bosons of which one intends to identify with the GBs of the Lorentz breaking.

The idea that gauge particles (photons and gravitons, for example) might arise as the GBs of a theory with spontaneous LSB has been widely studied and goes back a long way [1, 2, 3, 4]. Bumblebee models [5], which are vector theories exhibiting spontaneous LSB, have been useful in providing simple toy models to deal with this question [6, 7, 8, 9]. Also the study of Nambu models [10], where the LSB is imposed non-linearly in analogy with the description of pion interactions in the non-linear sigma model, brought new insights in this direction [11, 12, 13]. For additional references see for example Ref. [14].

We consider the simplest cases of the Abelian bumblebee and Nambu models which MGT is standard electrodynamics (ED). Nevertheless the method we use to determine the conditions for the equivalence between the Abelian Nambu model (ANM) and ED can be easily generalized to the non-Abelian case. The Abelian bumblebee model (ABM) and the ANM arise from the following Lagrangian density

$$\mathcal{L}_{\rm ABM} = -\frac{1}{4}B_{\mu\nu}B^{\mu\nu} - J_\mu B^\mu - \frac{\kappa}{2}V(\xi), \quad B_{\mu\nu} = \partial_\mu B_\nu - \partial_\nu B_\mu, \quad \xi = B_\mu B^\mu - n_\mu n^\mu b^2, \qquad (1)$$

and the difference between them depends on the rôle played by κ together with a specific choice of $V(\xi)$. In the former case κ is a constant with the usual choice $V(\xi) = \xi^n/n$, while in the latter situation κ is a Lagrange multiplier and $V(\xi) = \xi$. In Eq. (1), $b^2 > 0$, n^μ is a unit vector which indicates the direction of the symmetry breaking vacuum and the potential $V(\xi)$ has a minimum at $\xi = 0$, i.e. $V'(\xi)|_{\xi=0} = 0$ and $V''(\xi)|_{\xi=0} > 0$. The potential term breaks Lorentz symmetry spontaneously as well as gauge invariance. While bumblebee models contain massless and massive modes, which nature depends on the choice of the symmetry breaking vacuum, the constraint imposed on Nambu models via the Lagrange multiplier freezes the massive modes and restricts the dynamics to the Goldstone modes only.

The paper is organized as follows: in section 2 we review the conditions under which the ANM becomes equivalent to ED. Some comments regarding further generalizations of the method are included. Section 3 is devoted to a recap of the formulation of electrodynamics in terms of longitudinal and transverse variables. Here we pay careful attention to the use of a consistent gauge-fixing procedure at the classical level in electrodynamics, which clearly shows that the final Hamiltonian an Dirac brackets in terms of the transverse variables are fully independent of the gauge fixing. In this way we clarify some issues of gauge fixing arising in Ref. [15]. These results are subsequently employed in section 4 to obtain the phase space sector where the classical ABM is equivalent to ED. Our metric signature is $(+,-,-,-)$.

2. The Abelian Nambu model

In general, Nambu models can be derived from ABM models by taking κ to be a Lagrange multiplier and choosing $V(\xi) = \xi$. In practice, this means that the Lagrangian of the ANM is

$$\mathcal{L}_{\text{ANM}} = -\frac{1}{4}B_{\mu\nu}B^{\mu\nu} - J_\mu B^\mu, \qquad B_\mu B^\mu - n_\mu n^\mu b^2 = 0, \tag{2}$$

where the constraint must be solved and substituted in $\mathcal{L}_{\text{ANM}}$. Thus, the ANM has three degrees of freedom (d.o.f.), instead of the two required by gauge invariance in ED. We assume also that the ANM is coupled to a conserved current J_μ.

The problem now is how to recover gauge invariance from the ANM in a consistent manner. To this end, the main idea is to take advantage of the formal identity between the Lagrangians of the two theories, which will allow us to relate the corresponding canonical formulations through a canonical transformation, once we have identified the d.o.f. in the ANM which solve the constraint in Eq. (1). Let us consider the following independent coordinates Φ_3 and $\Phi_{\bar{k}}$, $\bar{k} = 1, 2$, together with the invertible velocity-independent transformation $B_i = B_i(\Phi_j)$

$$B_{\bar{k}} = \Phi_{\bar{k}}, \qquad B_3 = \Phi_3\left(1 - \frac{N}{4\Phi_3^2}\right), \qquad N = (\Phi_{\bar{k}}\Phi_{\bar{k}} + n^2 b^2). \tag{3}$$

The additional choice

$$B_0 = \Phi_3\left(1 + \frac{N}{4\Phi_3^2}\right), \tag{4}$$

guarantees the $B_\mu B^\mu = n^2 b^2$ is identically satisfied.

The above transformations (3) and (4) determine the velocities $\dot{B}_\mu$ as functions of $\dot{\Phi}_i$ and Φ_i and we can write $\mathcal{L}_{\text{ANM}} = \mathcal{L}_{\text{ANM}}(\Phi, \dot{\Phi})$. The calculation of the canonical momenta $\Pi = \partial\mathcal{L}_{\text{ANM}}/\partial\dot{\Phi}$ of the ANM indicates that we end up with a regular systems (no constraints) such that the velocities $\dot{\Phi}$ can be inverted in terms of the momenta Π. In other words, we can identify the three fields Φ_i as the d.o.f. of the ANM. The canonically-conjugated variables Φ and Π satisfy the standard Poisson algebra.

IOP Conf. Series: Journal of Physics: Conf. Series **952** (2018) 012013 doi:10.1088/1742-6596/952/1/012013

To make direct contact with ED, let us recall that $\mathcal{L}_{\rm ANM}$ in Eq. (2) can be rewritten as

$$\mathcal{L}_{\rm ANM}(\Phi,\dot{\Phi}) = \frac{1}{2}(e_i e_i - b_i b_i) - J^\mu B_\mu, \tag{5}$$

in terms of the standard electric $e_i = \dot{B}_i - \partial_i B_0$ and magnetic field $b_i = \epsilon_{ijk} B_{jk}/2$, which now are functions of $(\dot{\Phi},\Phi)$ and Φ, respectively. At this stage, e_i and b_i are only convenient labels to denote functions of Φ and Π.

Since (3) is a coordinate transformation we have

$$\frac{\partial \dot{B}_i}{\partial \dot{\Phi}_j} = \frac{\partial B_i}{\partial \Phi_j} \qquad \Rightarrow \qquad \Pi_i = \frac{\partial \mathcal{L}_{\rm ANM}}{\partial \dot{\Phi}_i} = \frac{\partial B_j}{\partial \Phi_i} e_j, \quad e_i(\Phi,\Pi) = \frac{\partial \Phi_j}{\partial B_i}\Pi_j. \tag{6}$$

Using the transformation $(\Phi_i,\Pi_j) \to (B_i, e_j)$ we can write

$$\mathcal{H}_{\rm ANM}(\Phi,\Pi) = \Pi_i\dot{\Phi}_i - \mathcal{L}_{\rm ANM} = \frac{1}{2}\left(e_i e_i + b_i b_i\right) + J^i B_i - \left(\partial_i e_i - J^0\right) B_0, \tag{7}$$

where $B_0 = B_3\left[(B_{\bar{i}}B_{\bar{i}} + n^2M^2)/(B_3B_3)\right]^{1/2}$ in terms of the original variables. Inserting the relation

$$\Pi_i\dot{\Phi}_i = \Pi_i \frac{\partial \Phi_i}{\partial B_j}\dot{B}_j = e_j \dot{B}_j. \tag{8}$$

in the Hamiltonian action of the ANM yields

$$S_{\rm ANM} = \int d^4x \left(\Pi_i\dot{\Phi}_i - \mathcal{H}_{\rm ANM}(\Phi,\Pi)\right) = \int d^4x \left(e^a_i \dot{B}^a_i - \mathcal{H}_{\rm ANM}(B_j, e_k)\right), \tag{9}$$

which identifies e_i as the canonically-conjugated momenta to B_i.

The transformation $(\Phi^a_i, \Pi^b_j) \to (B^a_i, e^b_j)$ is generated only by the coordinate transformation (3), so that it is a canonical transformation. Thus, without any further calculation, we recover the canonical algebra of ED among the variables B_i and e_j from that of the ANM. In fact, using the relations (3) together with (6) we can explicitly verify this statement by using the canonical algebra of the ANM.

Now we undo the canonical transformation in (9) and look at the quantities B_i and e_j as independent canonical variables satisfying such a Poisson algebra. Also the Hamiltonian density (7) acquires now the same form as in ED. Nevertheless, in spite of this apparent identity, $\mathcal{H}_{\rm ANM}$ is still not the Hamiltonian of ED because we are missing the following two conditions: (1) $\Omega \equiv (\partial_i e_i - J^0)$ is not a constraint and (2) B_0 is not an arbitrary function. To recover these conditions we calculate the evolution of the Gauss function Ω. We can do the calculation in the dynamics of the ANM using $\mathcal{H}_{\rm ANM}(\Phi,\Pi)$ together with the relations (3) and (6) or, in a simpler way, by considering directly $\mathcal{H}_{\rm ANM}(B_i, e_j)$ plus the associated canonical algebra we have just obtained. The result is

$$\dot{\Omega} = -\partial_\mu J^\mu + F\Omega + \partial_i(G^i\Omega), \tag{10}$$

where the functions F, G^i depend only on the fields B_j. Since we have assumed current conservation, we observe that Eq. (10) implies that after demanding $\Omega = 0$ at some initial time, we obtain the Gauss-law constraint for all time. Then we can add this constraint to the Hamiltonian density in the form $\Theta\,\Omega$, with Θ being an arbitrary function, which in turn will make $(B_0 + \Theta)$ arbitrary. Since Ω is now a constraint, Eq. (10) assures us that no additional constraints arise.

Summarizing: the equivalence between the ANM and the ED, both coupled to a conserved external current, is obtained only after the Gauss law is imposed as an initial condition. The above procedure can be directly generalized for a non-Abelian Nambu model when finding the conditions for its equivalence with the corresponding non-Abelian MGT [16].

We have also provided a proof of the quantum equivalence of the ANM and ED in the gauge $B_\mu B^\mu - n^2 b^2 = 0$, to all orders in perturbation theory, under the previously stated requirements. In this case the ANM is coupled to the conserved fermionic current $e\bar{\Psi}\gamma^\mu\Psi$ and the Gauss law is imposed upon the zeroth-order Lagrangian which reduces to ED in the axial gauge for $t \to \infty$ [17].

The study of Nambu models poses the following general problem in gauge theories: how do we recover gauge invariance after breaking it by the imposition of an additional constraint among the coordinates? This question was addressed in Ref. [18] for a Yang-Mills theory. We dealt with this problem in Ref. [14] by introducing what we call an extended Nambu model, defined by a Lagrangian density yielding first-class constraints only, plus an arbitrary relation among the coordinates restricted alone by very general conditions.

3. Electrodynamics in transverse and longitudinal variables

In preparation for the next section, let us briefly recall the formulation of ED in terms of these variables, which clearly shows that the resulting Hamiltonian and Dirac algebra are independent of the gauge fixing. To this end we decompose any vector field V^k in its transverse V_T^k and longitudinal $V_L^k = \partial_k V$ parts, such that $V^k = V_T^k + \partial_k V$, satisfying $\nabla \cdot \mathbf{V}^T = 0$ and $\nabla \times \mathbf{V}^L = 0$. In terms of the original vector we have

$$V_T^i = \left(\delta_{ik} - \frac{\partial_i \partial_k}{\nabla^2}\right) V^k, \qquad V = \frac{1}{\nabla^2}\partial_k V^k, \qquad \nabla^2 = \partial_k \partial_k. \tag{11}$$

For ED we rewrite the vector potential A^μ in terms of A_0, A_T^i, A and use analogous decomposition for the current J^μ. Using these variables, the standard Lagrangian density for ED becomes

$$\mathcal{L}_{ED} = \frac{1}{2}(\dot{A}_T^i)^2 - \frac{1}{2}\dot{A}\,\nabla^2 \dot{A} - (\nabla^2 A_0)\,\dot{A} - \frac{1}{2}A^0 \nabla^2 A^0 - \frac{1}{4}F_T^{ij}F_{Tij} - J^0 A^0 - J_T^i A_T^i - J\nabla^2 A. \tag{12}$$

The equations of motion are

$$-\ddot{A}_T^j + \partial_i F_T^{ij} + J_T^j = 0, \qquad \nabla^2(\dot{A} + A^0) + J^0 = 0, \qquad \nabla^2(\ddot{A} + \dot{A}_0 - J) = 0, \tag{13}$$

arising from the variations δA_T^i, δA_0 and δA respectively. Let us observe that current conservation $\partial_0 J^0 + \partial_i J^i = 0$ reads $\dot{J}^0 = -\nabla^2 J$ in such a way that the third equation in (13) is just the time derivative of the second equation.

The canonical momenta are

$$\Pi_0 = \frac{\partial L}{\partial \dot{A}^0} = 0, \quad \Pi_{iT} = \frac{\partial L}{\partial \dot{A}_T^i} = \dot{A}_T^i, \quad \Pi = \frac{\partial L}{\partial \dot{A}} = -\nabla^2(\dot{A} + A_0). \tag{14}$$

yielding the non-zero, equal-time Poisson brackets

$$\begin{aligned} \{A^0(t,\mathbf{x}),\, \Pi_0(t,\mathbf{y})\} &= \delta^3(\mathbf{x}-\mathbf{y}), \qquad \{A(t,\mathbf{x}),\, \Pi(t,\mathbf{y})\} = \delta^3(\mathbf{x}-\mathbf{y}), \\ \{A_T^i(t,\mathbf{x}),\, \Pi_{Tj}(t,\mathbf{y})\} &= \left(\delta_{ij} - \frac{\partial_i\partial_j}{\nabla^2}\right)\delta^3(\mathbf{x}-\mathbf{y}). \end{aligned} \tag{15}$$

The extended Hamiltonian is

$$\mathcal{H}_E = \frac{1}{2}(\Pi_{iT})^2 + \frac{1}{4}F_{Tij}F_T^{ij} - J_T^i A_T^i - \frac{1}{2}\Pi\frac{1}{\nabla^2}\Pi + (J^0 - \Pi)A_0 + J\nabla^2 A + \alpha\Pi_0, \tag{16}$$

IOP Conf. Series: Journal of Physics: Conf. Series **952** (2018) 012013 doi:10.1088/1742-6596/952/1/012013

where we have explicitly introduced the primary constraint $\Pi_0 \approx 0$. The first-order action leading to the Hamilton equations reads

$$S = \int d^4x \left(\Pi_0 \dot{A}_0 + \Pi_{iT} \dot{A}^i_T + \Pi \dot{A} - \mathcal{H}_E \right), \tag{17}$$

Applying the standard Dirac procedure we are finally left with two first-class constraints

$$\phi_1 = \Pi_0 \approx 0, \qquad \phi_2 = \Pi - J^0 \approx 0\,, \tag{18}$$

which lead to the final count of two independent d.o.f. in coordinate space. Let us observe that current conservation allows to rewrite the following two contributions in the Hamiltonian action as

$$\int d^4x(\Pi\dot{A} - J\nabla^2 A) = \int d^4x(\Pi\dot{A} + J^i\partial_i A) = \int d^4x(\Pi\dot{A} - \partial_i J^i A) = \int d^4x(\Pi - J^0)\dot{A} \approx 0. \tag{19}$$

In this way, after the complete gauge fixing, that is to say after the constraints (18) have been set strongly equal to zero, these two pieces will always cancel, irrespectively of the explicit form of the additional functions chosen for the gauge fixing. Also the contribution proportional to $(J^0 - \Pi)$ in the Hamiltonian density will be zero. Thus, after imposing (18) strongly, which means having fixed the gauge via two additional constraints, the Hamiltonian action always reduces to

$$S = \int d^3x \left(\Pi_{iT}\dot{A}^i_T - \frac{1}{2}(\Pi_{iT})^2 - \frac{1}{4}F_{Tij}F^{ij}_T + J^i_T A^i_T + \frac{1}{2}J^0 \frac{1}{\nabla^2} J^0 \right). \tag{20}$$

An important point to have in mind for a correct gauge fixing is to verify that the following conditions are satisfied. (a) The consistency of the additional constraints with the equations of motion, that is to say that the proposed choices of A_0 and A are compatible with the second equation in (13). (b) That the resulting system is in fact second class. (c) That the resulting Dirac brackets of the remaining transverse variables are unchanged with respect to their PB.

As a first general step, we can always use the constraint $\phi_1 = \Pi_0 \approx 0$ together with

$$\phi_3 = A_0 + \dot{A} + \frac{1}{\nabla^2} J^0, \tag{21}$$

which is precisely the second equation in (13), as a set of second class constraints to get rid of the variables Π_0 and A^0. This choice maintains the Dirac brackets of the transverse degrees of freedom unchanged with respect to the original brackets in Eq. (15). Finally one has to set the constraint ϕ_2 strongly equal to zero. We illustrate this last gauge fixing by imposing a general coordinate constraint

$$\phi_4 = F(A^\mu) \approx 0. \tag{22}$$

In the variables we are using this means to set $F(A^0 = -\partial_t A - \frac{1}{\nabla^2}J^0, A^k_T, \partial_k A) \approx 0$, which defines a complicated non-linear equation for $A = A(A^k, J_0)$. Adopting a solution $Q(A^k, J_0)$ of this equation we replace ϕ_4 by

$$\bar{\phi}_4 = A - Q(A^k, J_0) \approx 0, \tag{23}$$

which is such that the set $\{\bar{\phi}_4, \phi_2\}$ is second class, thus fixing completely the remaining gauge symmetry. In practice, the explicit form of Q can be obtained only in very particular cases, but we assume its existence [19]. Again, the Dirac brackets of the transverse variables remain unchanged and the final Hamiltonian action is given by Eq. (20).

IOP Conf. Series: Journal of Physics: Conf. Series **952** (2018) 012013 doi:10.1088/1742-6596/952/1/012013

4. The Abelian bumblebee model

In this section we present a proof of the equivalence between the classical ABM and ED, under conditions to be specified in the following, which is alternative to that given in Ref. [15]. Let us recall that κ in Eq. (1) is now a constant instead of a Lagrange multiplier, which makes this theory different from the ANM discussed in the previous sections. To identify the phase space sector where the equivalence is achieved we follow closely the formulation of ED in terms of longitudinal and transverse variables discussed in section 3.

Since $\mathcal{L}_{ABM}(B^\mu) = \mathcal{L}_{ED}(B^\mu) - \frac{\kappa}{2}V(\xi)$, the equations of motion are

$$-\ddot{B}^j_T + \partial_i B^{ij}_T + J^j_T - \kappa V'(\xi)B_j = 0, \qquad \nabla^2(\dot{B} + B^0) + J^0 - \kappa V'(\xi)B^0 = 0,$$
$$\nabla^2(\ddot{B} + \dot{B}_0 - J) - \kappa\partial_i\left(V'(\xi)B^i\right) = 0. \tag{24}$$

We always keep in mind that $B^i = B^i_T + \partial_i B$ but for notational simplicity we use the compact notation B^i (or B^μ), unless confusion arises.

Because the potential $V(\xi)$ does not include velocities, the canonically conjugated momenta of the ABM, together with the corresponding Poisson brackets are just given by the expressions (14) and (15). Also, the Hamiltonian density is $\mathcal{H}_{ABM} = \mathcal{H}_{ED} + \frac{\kappa}{2}V(\xi)$. The ABM has only two constraints

$$\Phi_1 = \Pi_0 \approx 0, \qquad \Phi_2 = \Pi - J^0 - \kappa V'(\xi)B^0 \approx 0, \tag{25}$$

which are second class, implying that we have $\frac{1}{2}(2\times 4-2) = 3$ d.o.f., instead of the two arising in ED. One possibility to cut this additional d.o.f. is to impose two additional second-class constraints.

In order to come closely to ED we propose to incorporate the additional constraints

$$\Phi_3 = B^0 + \dot{B} + \frac{1}{\nabla^2}J^0 \approx 0, \qquad \Phi_4 = B_\mu B^\mu - n^2b^2 = \xi \approx 0. \tag{26}$$

Nevertheless, in order to keep a sound theory we have to determine under which conditions such additional constraints are maintained for all time by the dynamics of the ABM, so that no further constraints are generated. We begin by calculating $\dot{\Phi}_3$ from Eq. (26), which involves $\dot{B}^0 + \ddot{B}$. Solving for this combination in the last equation in (24) yields

$$\dot{\Phi}_3 = \frac{1}{\nabla^2}\Big(\partial_\mu J^\mu + \kappa\partial_i\big(V'(\xi)B^i\big)\Big) = \frac{1}{\nabla^2}\Big(\partial_\mu J^\mu + \kappa V'(\xi)\partial_i B^i + \kappa V''(\xi)B^i\partial_i(B_\mu B^\mu)\Big). \tag{27}$$

From the above relation we conclude that after imposing current conservation and the condition $B_\mu B^\mu = n^2b^2$ ($\xi = 0$, $\partial_i(B_\mu B^\mu) = 0$) at some initial time, $t = 0$ for example, the constraint $\Phi_3 \approx 0$ is conserved for all time.

The next step is to calculate $\dot{\Phi}_4 = 2(B_0\dot{B}_0 - B_i\dot{B}_i)$, which we do by solving for $\dot{B}^0$ from the explicit expression for $\dot{\Phi}_2 \approx 0$. We have

$$\dot{\phi}_2 = \dot{\Pi} - \dot{J}^0 - \kappa\left(\dot{B}_0 V' + 2B_0V''(B_0\dot{B}_0 - B_i\dot{B}_i)\right) = 0, \tag{28}$$

which leads to

$$\dot{B}_0 \approx \frac{1}{\kappa(V' + 2B_0^2V'')}\left(\dot{\Pi} - \dot{J}^0 + 2\kappa B_0V''B_i\dot{B}_i\right). \tag{29}$$

Recalling that $\Pi = -\nabla^2(\dot{B}+B_0)$, leading to $\dot{\Pi} = -\nabla^2(\ddot{B}+\dot{B}_0)$, and using again the full equation arising from the third expression in the relations (24) we obtain $\dot{\Pi} = -\partial_i J^i - \kappa\partial_i[V'(\xi)B^i]$ which we substitute in (29), yielding the final expression

$$\dot{B}_0 \approx \frac{1}{\kappa(V' + 2B_0^2V'')}\left(-\partial_\mu J^\mu - \kappa\partial_i\big(V'(\xi)B^i\big) + 2\kappa B_0V''B_i\dot{B}_i\right). \tag{30}$$

The final step is to replace $\dot{B}_0$ in the expression for $\dot{\Phi}_4$ which leads to

$$\dot{\Phi}_4 \approx \frac{2B_0}{\kappa(V' + 2B_0^2 V'')} \left(-\partial_\mu J^\mu - \kappa \partial_i \big(V'(\xi) B^i\big) - \frac{\kappa V'(\xi)}{B_0} B_i \dot{B}_i \right). \tag{31}$$

Again, after imposing the initial conditions $\partial_\mu J^\mu|_{t=0} = 0$ and $\xi|_{t=0} = (B_\mu B^\mu - n^2 b^2)|_{t=0} = 0$, we obtain $\dot{\Phi}_4 \approx 0$ for all time.

To conclude, let us summarize the phase-space conditions under which the ABM turns into ED. As mentioned before, we set the initial conditions $\partial_\mu J^\mu|_{t=0} = 0$ and $\xi|_{t=0} = (B_\mu B^\mu - n^2 b^2)|_{t=0} = 0$. Then, at this time we recover the set of constraints (18), (21) and (22) given in the previous section, the latter with the choice $F(B^\mu) = B_\mu B^\mu - n^2 b^2$. Also, up to the constant $\kappa V(0)/2$, we have $\mathcal{H}_{ABM} = \mathcal{H}_{ED}$. That is to say, at $t = 0$ the above conditions turn the ABM into ED in the gauge $B_\mu B^\mu - n^2 b^2 = 0$. The constraints Φ_1 and Φ_2 in the ABM are automatically conserved by their own dynamics, and the above initial conditions guarantee also that $\dot{\Phi}_3 \approx 0$ and $\dot{\Phi}_4 \approx 0$ at $t = 0$. Then we reproduce the constraints (18), (21) and (22) together with $\mathcal{H}_{ED}$, now at $t = 0 + \delta t$, which means that the ABM is equivalent to ED at $t = 0 + \delta t$ in this sector of the phase space. We can iterate the previous analysis, since having Maxwell equations at $t = 0 + \delta t$, the identity $0 \equiv \partial_\mu \partial_\nu B^{\mu\nu} = \partial_\mu J^\mu$, leads to current conservation at this later time. In other words, the two previously defined initial conditions determine the sector of the phase space where the classical ABM becomes equivalent to ED for all time. A proof of the quantum equivalence between the two models to all orders in perturbation theory is given in Ref. [15], following similar steps to those in Ref. [17] for the ANM.

Acknowledgments

LFU thanks the Organizing Committee of the International Workshop on CPT and Lorentz Symmetry in Field Theory for a wonderful meeting in Faro. LFU has been partially supported by the project CONACyT # 237503. CAE and LFU acknowledge support from the project UNAM (Dirección General de Asuntos del Personal Académico) # IN104815. CAE is supported by the CONACyT Posdoctoral Grant No. 234745.

References

[1] Bjorken J D 1963 *Ann. Phys. (N.Y.)* **24** 174
[2] Guralnik G S 1964 *Phys. Rev.* **136** B1404
[3] Guralnik G S 1964 *Phys. Rev. Letter* **13** 295
[4] Ivanov E A and Ogievetsky V I 1976 *Lett. Math. Phys.* **1** 309
[5] Kostelecký V A and Samuel S 1989 *Phys. Rev.* D **40** 1886
[6] Bluhm R, Fung S H and Kostelecký V A 2008 *Phys. Rev.* D **77** 065020
[7] Bluhm R, Gagne N L, Potting R and Vrublevskis A 2008 *Phys. Rev.* D **77** 125007
[8] Kostelecký V A and Potting R 2009 *Phys. Rev.* D **79** 065018
[9] Carroll S M, Tam H and Wehus I K 2009 *Phys. Rev.* D **80** 025020
[10] Nambu Y 1968 *Progr. Theor. Phys. Suppl. Extra* 190
[11] Azatov A T and Chkareuli J L 2006 *Phys. Rev.* D **73** 065026
[12] Chkareuli J L, Froggatt C D and Nielsen H B 2009 *Nucl. Phys.* B **821** 65
[13] Chkareuli J L, Jejelava J G and Tatishvili G 2011 *Phys. Lett.* B **696** 124
[14] Escobar C A and Urrutia L F 2017 *Int. J. Mod. Phys.* A **32** 1750077
[15] Martín-Ruiz A and Escobar C A 2017 *Phys. Rev.* D **95** 095006
[16] Escobar C A and Urrutia L F 2015 *Phys. Rev.* D **92** 025013
[17] Escobar C A and Urrutia L F 2015 *Phys. Rev.* D **92** 025042
[18] Jackiw R 1980 *Rev. Mod. Phys.* **52** 661
[19] Franca O J, Montemayor R and Urrutia L F 2012 *Phys. Rev.* D **85** 085008

IOP Conf. Series: Journal of Physics: Conf. Series **952** (2018) 012014 doi:10.1088/1742-6596/952/1/012014

Constraining spacetime nonmetricity with Lorentz-violation methods

Zhi Xiao[1,2], Ralf Lehnert[2,3], W M Snow[2,4,5] and Rui Xu[4]

[1] North China Electric Power University, Beijing 102206, China
[2] Indiana University Center for Spacetime Symmetries, Bloomington, IN 47405, USA
[3] Leibniz Universität Hannover, Welfengarten 1, 30167 Hannover, Germany
[4] Physics Department, Indiana University, Bloomington, IN 47405, USA
[5] Center for Exploration of Energy and Matter, Indiana University, Bloomington, IN 47408, USA

Abstract. In this report, we will give the first constraints on in-matter nonmetricity. We will show how the effective-field-theory (EFT) toolbox developed for the study of Lorentz violation (LV) can be employed for investigations of the "effective LV" background caused by nonmetricity, a geometric object extending the notion of a Riemannian manifold. The idea is to probe for the effects of spacetime nonmetricity sourced by liquid ^{4}He with polarized slow neutrons. We present the first constraints on isotropic and parity-odd nonmetricity components. Further constraints on anisotropic nonmetricity components within this EFT framework may be feasible with proper experimental techniques in the near future.

1. Introduction

The celebrated observation in 2016 of gravitational waves by the LIGO and Virgo Scientific Collaboration [1] has once more underscored the widely held belief that the geometry of spacetime is a dynamical physical entity. Keeping such a geometric foundation, efforts to extend general relativity (GR) proceed often by exploring more general geometries beyond that of a Riemanian spacetime. In general metric-affine gravity [2], the linear connection is $\Gamma^{\alpha}{}_{\beta\gamma} = \{{}_{\beta}{}^{\alpha}{}_{\gamma}\} + K^{\alpha}{}_{\beta\gamma} + Q^{\alpha}{}_{\beta\gamma}$, where $\{{}_{\beta}{}^{\alpha}{}_{\gamma}\}$ denotes the Levi–Civita connection determined purely by the metric tensor $g_{\mu\nu}$, $K^{\alpha}{}_{\beta\gamma} \equiv \frac{1}{2}\left(T^{\alpha}{}_{\beta\gamma} - T_{\beta}{}^{\alpha}{}_{\gamma} - T_{\gamma}{}^{\alpha}{}_{\beta}\right)$ is the contorsion tensor determined by the torsion tensor $T^{\alpha}{}_{\beta\gamma} \equiv \Gamma^{\alpha}{}_{\beta\gamma} - \Gamma^{\alpha}{}_{\gamma\beta}$, and $Q^{\alpha}_{\beta\gamma} \equiv \frac{1}{2}\left(N_{\beta}{}^{\alpha}{}_{\gamma} + N_{\gamma}{}^{\alpha}{}_{\beta} - N^{\alpha}{}_{\beta\gamma}\right)$ is given by the nonmetricity tensor $N_{\alpha\beta\gamma} \equiv -D_{\alpha}g_{\beta\gamma}$. In a historical context, two avenues for extending GR are particularly noteworthy. By relaxing the usual metric-compatibility condition $D_{\alpha}g_{\beta\gamma} = 0$, Weyl interpreted $(N_1)_{\alpha} = g^{\beta\gamma}N_{\alpha\beta\gamma}$ as the electromagnetic 4-potential in a (failed) attempt to unify electromagnetism with gravity [3]. By relaxing the torsion-free condition $\Gamma^{\alpha}{}_{\beta\gamma} = \Gamma^{\alpha}{}_{\gamma\beta}$ instead, Cartan proposed to couple torsion with the intrinsic-spin current of matter fields, an approach now known as Einstein–Cartan theory [4].

Assuming that $T^{\alpha}{}_{\beta\gamma}$ or $N_{\alpha\beta\gamma}$ vary slowly at the scale of the solar system, we may regard them as constant tensor fields in the neighborhood of the Earth. For a local terrestrial experiment, these tensor fields then effectively break Lorentz symmetry even if the original theory is Lorentz invariant [5]. This idea allows us to utilize the theoretical framework developed for the study of CPT- and Lorentz-symmetry violation, the Standard-Model Extension (SME) [6], to explore the associated background-tensor–matter couplings. For example, some characteristic LV signals, like sidereal and annual variations of physical observables [7], which stem from the motion of Earth-based laboratories through such solar torsion or nonmetricity backgrounds, may be used to constrain these tensor fields. This idea has already been studied extensively for torsion backgrounds [8] including in-matter torsion effects [9]. Numerous other phenomenological studies on torsion fields are also available in the literature [10–19]. For the case of nonmetricity–matter couplings, an analogous treatment has been given recently in Ref. [20]. As an important geometric object, matter–nonmetricity couplings have also been the focus of various theoretical studies [21]. However, in this paper we adopt the viewpoint of Ref. [20] and regard the presence of nonmetricity as an experimental issue instead.

In particular, we study in-matter nonmetricity couplings by considering a polarized slow-moving neutron beam propagating in liquid ^{4}He. In this set-up, the source of nonmetricity is assumed to be the liquid ^{4}He, and the probe is the neutron beam. The polarized neutrons are produced at the National Institute of Standards and Technology (NIST) Center for Neutron Research and have an approximate Maxwellian distribution with a peak energy of about 3 meV. The liquid ^{4}He was kept at a temperature of around 4 K in a magnetically shielded cryogenic target [22]. The key difference between our study and that in Ref. [20] is that we examine the hypothetical nonmetricity sourced locally by liquid ^{4}He in a terrestrial laboratory. Since the ^{4}He is comoving with the laboratory, this excludes any sidereal or annual effect predicted in studies of nonmetricity sourced by celestial body, *e.g.*, the Sun.

This paper is organized as follows. Section 2 briefly reviews the effective-field-theory framework borrowed from LV studies to describe fermion–nonmetricity couplings. In Sec. 3, we discuss the constraints obtained from the measurement of neutron-spin precession at NIST's Center for Neutron Research. Finally, in Sec. 4, we briefly summarize and discuss the results of this paper. The convention for the metric signature and the Levi–Civita symbol are $\eta^{\mu\nu} = \mathrm{diag}(+,-,-,-)$ and $\epsilon^{0123} = +1$, respectively.

2. Theoretical Preparation

Since we adopt an experimental point of view, as mentioned above, the question as to whether $N_{\alpha\beta\gamma}$ should be minimally coupled (to the dilation and shear currents) [21] or not is not our primary concern. Our starting point is the analysis in Ref. [20] that takes all matter–nonmetricity operators allowed by observer Lorentz symmetry [6] into account. To our knowledge, $N_{\alpha\beta\gamma}$ must be tiny, which means we can safely drop higher-order nonmetricity couplings. The relevant energy in the neutron experiment is comparatively low, which means we may also ignore the composite nature of the neutron and regard it as a point-like particle. The free neutron Lagrangian is $\mathcal{L}_0 = \frac{i}{2}\overline{\psi}\gamma^\mu \overleftrightarrow{\partial}_\mu \psi - m\overline{\psi}\psi$. We also set $g^{\mu\nu} \to \eta^{\mu\nu}$ since we may ignore conventional gravitational effect for our current purposes. To write down systematically fermion–nonmetricity couplings in the SME framework, it will be convenient to decompose $N_{\alpha\beta\gamma}$ into its four Lorentz irreducible pieces [20], *i.e.*,

$$\begin{aligned}
(N_1)_\mu \equiv& -\eta^{\alpha\beta}N_{\mu\alpha\beta}, \qquad (N_2)_\mu \equiv -\eta^{\alpha\beta}N_{\alpha\mu\beta},\\
S_{\mu\alpha\beta} \equiv& \tfrac{1}{3}\big[N_{\mu\alpha\beta}+N_{\alpha\beta\mu}+N_{\beta\mu\alpha}\big] + \tfrac{1}{18}\big[(N_1)_\mu\,\eta_{\alpha\beta}+(N_1)_\alpha\,\eta_{\beta\mu}+(N_1)_\beta\,\eta_{\mu\alpha}\big]\\
&+\tfrac{1}{9}\big[(N_2)_\mu\,\eta_{\alpha\beta}+(N_2)_\alpha\,\eta_{\beta\mu}+(N_2)_\beta\,\eta_{\mu\alpha}\big],\\
M_{\mu\alpha\beta} \equiv& \tfrac{1}{3}\big[2N_{\mu\alpha\beta}-N_{\alpha\beta\mu}-N_{\beta\mu\alpha}\big] + \tfrac{1}{9}\big[2(N_1)_\mu\,\eta_{\alpha\beta}-(N_1)_\alpha\,\eta_{\beta\mu}-(N_1)_\beta\,\eta_{\alpha\mu}\big]\\
&-\tfrac{1}{9}\big[2(N_2)_\mu\,\eta_{\alpha\beta}-(N_2)_\alpha\,\eta_{\beta\mu}-(N_2)_\beta\,\eta_{\alpha\mu}\big].
\end{aligned} \tag{1}$$

The nonmetricity tensor can be rebuilt from the above Lorentz irreducible tensors as follows

$$\begin{aligned}
N_{\mu\alpha\beta} =& \tfrac{1}{18}\big[-5(N_1)_\mu\eta_{\alpha\beta}+(N_1)_\alpha\eta_{\beta\mu}+(N_1)_\beta\eta_{\mu\alpha}+2(N_2)_\mu\eta_{\alpha\beta}-4(N_2)_\alpha\eta_{\beta\mu}-4(N_2)_\beta\eta_{\mu\alpha}\big]\\
&+S_{\mu\alpha\beta}+M_{\mu\alpha\beta}\,.
\end{aligned} \tag{2}$$

With these Lorentz-irreducible pieces, we can list all the relevant Lagrangian contributions constructed first in Ref. [20]:

$$\begin{aligned}
\mathcal{L}_N^{(4)} &= \left[\zeta_1^{(4)}\,(N_1)_\mu + \zeta_3^{(4)}\,(N_2)_\mu\right]\overline{\psi}\gamma^\mu\psi + \left[\zeta_2^{(4)}\,(N_1)_\mu + \zeta_4^{(4)}\,(N_2)_\mu\right]\overline{\psi}\gamma_5\gamma^\mu\psi,\\
\mathcal{L}_N^{(5)} &= -\tfrac{1}{2}\left\{i\left[\zeta_1^{(5)}\,(N_1)^\mu + \zeta_3^{(5)}\,(N_2)^\mu\right]\overline{\psi}\overleftrightarrow{\partial}_\mu\psi + \left[\zeta_2^{(5)}\,(N_1)^\mu + \zeta_4^{(5)}\,(N_2)^\mu\right]\overline{\psi}\gamma_5\overleftrightarrow{\partial}_\mu\psi\right\}\\
&\quad - \tfrac{1}{4}i\left[\zeta_5^{(5)}\,M_{\mu\nu}{}^\rho - \zeta_6^{(5)}\,{}^*M_{\mu\nu}{}^\rho\right]\overline{\psi}\sigma^{\mu\nu}\overleftrightarrow{\partial}_\rho\psi + \tfrac{1}{2}i\left[\zeta_7^{(5)}\,(N_1)_\mu + \zeta_8^{(5)}\,(N_2)_\mu\right]\overline{\psi}\sigma^{\mu\nu}\overleftrightarrow{\partial}_\nu\psi\\
&\quad - \tfrac{1}{4}i\,\epsilon^{\lambda\mu\nu\rho}\left[\zeta_9^{(5)}\,(N_1)_\lambda + \zeta_{10}^{(5)}\,(N_2)_\lambda\right]\overline{\psi}\sigma_{\mu\nu}\overleftrightarrow{\partial}_\rho\psi,\\
\mathcal{L}_N^{(6)} &\supset -\tfrac{1}{4}\zeta_1^{(6)}\,S_\lambda{}^{\mu\nu}\,\overline{\psi}\gamma^\lambda\partial_\mu\partial_\nu\psi + \text{h.c.} - \tfrac{1}{4}\zeta_2^{(6)}\,S_\lambda{}^{\mu\nu}\,\overline{\psi}\gamma_5\gamma^\lambda\partial_\mu\partial_\nu\psi + \text{h.c.}\,,
\end{aligned}\tag{3}$$

where we have defined ${}^*M_{\mu\nu}{}^\rho \equiv \frac{1}{2}\epsilon_{\kappa\lambda\mu\nu}\,M^{\kappa\lambda\rho}$ and all $\zeta_l^{(n)}$ are real. To avoid missing any possible nonmetricity–matter couplings, we have to take all Lorentz irreducible pieces of $N_{\mu\alpha\beta}$ into account including $M_{\alpha\beta\gamma}$ and $S_{\alpha\beta\gamma}$ in $\mathcal{L}_N^{(5)}$, $\mathcal{L}_N^{(6)}$. From the study of Lorentz violation [6, 23, 24], we know that some LV coefficients can be eliminated by field redefinitions, and a similar situation arises for some of the above nonmetricity operators. By scrutinizing Lagrangian (3), the operators associated with the couplings $\zeta_1^{(4)}$, $\zeta_3^{(4)}$, $\zeta_1^{(5)}$, $\zeta_2^{(5)}$, $\zeta_3^{(5)}$, $\zeta_4^{(5)}$, $\zeta_7^{(5)}$, and $\zeta_8^{(5)}$ can be removed at least at first order with carefully chosen field redefinitions.

By matching to our concrete experimental set-up, we can refine the remaining Lagrangian further. First note that the ground state of ^{4}He is spin 0. This state can therefore not contribute to anisotropies. Although individual ^{4}He atoms can have specific positions and momenta, no particular arrangement of a large numbers of helium atoms with preferred directions is expected. So, on average, nonmetricity generated by a macroscopic number of ^{4}He atoms is expected to be rotationally invariant, which means we only need to consider isotropic couplings. Components without spatial indices are obviously isotropic, so we have $(N_1)^0$, $(N_2)^0$, and S^{000}. For the mixed-symmetry piece, note that it satisfies the "Bianchi-like" identity

$$M_{\alpha\beta\gamma} + M_{\beta\gamma\alpha} + M_{\gamma\alpha\beta} = 0\,, \tag{4}$$

which indicates that no mixed-symmetry parts contributes. For a more detailed explanation, see Ref. [25]. Next, note that a more suitable framework to deal with slow-neutron effects is nonrelativistic. To extract the nonrelativistic physics we found it convenient to utilize directly a result in Ref. [24] instead of resorting to the usual Foldy–Wouthuysen transformation [26]. The final nonrelativistic Hamiltonian becomes

$$h_s = \frac{\vec{p}^2}{2m} + \left[\left(\zeta_2^{(4)} - m\,\zeta_9^{(5)}\right)(N_1)_0 + \left(\zeta_4^{(4)} - m\,\zeta_{10}^{(5)}\right)(N_2)_0 + \frac{m^2}{2}\zeta_2^{(6)}S_{000}\right]\frac{\vec{p}\cdot\vec{\sigma}}{m}\,, \tag{5}$$

where we have ignored the rest-mass contribution. Note that the term with the square brackets is spin dependent and can be traced back to a parity-odd, time-reversal even operator. A more complete nonrelativistic Hamiltonian including anisotropic contributions can be found in Ref. [25].

3. Experiment and Constraint

As mentioned above, the nonmetricity–neutron coupling operator in Eq. (5) is parity violating (PV). When a transversely polarized slow-neutron beam propagates through such a nonmetricity background, its spin vector rotates. To see this, we give a brief heuristic derivation below. For a beam of neutrons propagating in the z direction with its spin transversely polarized in a direction $\vec{n} = (\cos\alpha, \sin\alpha, 0)$, the initial state can be written as $|\psi_\perp\rangle = \frac{1}{\sqrt{2}}(e^{-i\alpha/2}|+\rangle + e^{i\alpha/2}|-\rangle)$, where

$|\pm\rangle$ represent helicity eigenstates. After passing a distance $L = |\vec{p}|t/m$ through liquid ^{4}He, the presumed source of nonmetricity, the Hamiltonian (5) indicates that, up to a total phase, $|\psi_\perp\rangle$ evolves to

$$|\psi_\perp(t)\rangle = \frac{1}{\sqrt{2}}(e^{-i(\alpha+\phi_{PV})/2}|+\rangle + e^{i(\alpha+\phi_{PV})/2}|-\rangle)\,, \tag{6}$$

where

$$\phi_{PV} = 2\Big[\Big(\zeta_2^{(4)} - m\,\zeta_9^{(5)}\Big)(N_1)_0 + \Big(\zeta_4^{(4)} - m\,\zeta_{10}^{(5)}\Big)(N_2)_0 + \frac{m^2}{2}\zeta_2^{(6)}S_{000}\Big]L\,. \tag{7}$$

Comparing to the initial state $|\psi_\perp\rangle$, ϕ_{PV} is just the spin rotation angle. In neutron physics, this phenomena is called neutron optical activity and is quantified by the rotary power $\frac{d\phi_{PV}}{dL}$. With an ingenious experimental design to cancel residual magnetic effects and other common-mode signals, this quantity was measured at NIST's Center for Neutron Research at the 1-σ level:

$$\frac{d\phi_{PV}}{dL} = +1.7 \pm 9.1(\text{stat.}) \pm 1.4(\text{sys}) \times 10^{-7}\,\text{rad/m}\,. \tag{8}$$

For details of this measurement, refer to Refs. [27–30]. Conversion to natural units together with Eq. (7) yields the following nonmetricity measurement:

$$\Big(\zeta_2^{(4)} - m\,\zeta_9^{(5)}\Big)(N_1)_0 + 2\Big(\zeta_4^{(4)} - m\,\zeta_{10}^{(5)}\Big)(N_2)_0 + m^2\,\zeta_2^{(6)}S_{000} = (3.4 \pm 18.2) \times 10^{-23}\,\text{GeV}\,. \tag{9}$$

We interpret this result as the 2-σ constraint

$$\Big|2\Big(\zeta_2^{(4)} - m\,\zeta_9^{(5)}\Big)(N_1)_0 + 2\Big(\zeta_4^{(4)} - m\,\zeta_{10}^{(5)}\Big)(N_2)_0 + m^2\,\zeta_2^{(6)}S_{000}\Big| < 3.6 \times 10^{-22}\,\text{GeV}\,. \tag{10}$$

Without extremely fine-tuned cancellations between the various nonmetricity couplings in Ref. (10), we may give the following individual bounds:

$$\begin{aligned} &|\zeta_2^{(4)}(N_1)_0| < 10^{-22}\,\text{GeV}, \quad |\zeta_4^{(4)}(N_2)_0| < 10^{-22}\,\text{GeV}\,, \\ &|\zeta_9^{(5)}(N_1)_0| < 10^{-22}, \quad |\zeta_{10}^{(5)}(N_2)_0| < 10^{-22}, \quad |\zeta_2^{(6)}S_{000}| < 10^{-22}\,\text{GeV}^{-1}. \end{aligned} \tag{11}$$

The above limits, to our knowledge, provide the first measurement of $\zeta_2^{(6)}S_{000}$ as well as the first measurement of any nonmetricity component inside matter.

4. Summary

In this work, using the SME and a measurement of neutron rotary power, we have obtained the constraint (10) on isotropic nonmetricity–neutron couplings. Without any fine-tuned cancellation, this constraint can be further interpreted as bounds on various nonmetricity components (11). Note that the well-known PV physics in the usual Standard Model (SM) also leads to neutron spin rotation, but no quantitative result from first-principle calculations is available for this conventional effect due to the intractability of the strong interactions. However, from the most decent estimate in the literature [31], $(d\phi_{PV}/dL)_{\text{SM}} = -6.5 \pm 2.2 \times 10^{-7}\,\text{rad/m}$, which is still less than the statistics of the measurement (8), we conclude that the constraint (10) we extracted on nonmetricity is reliable. The SM background prevents more stringent bounds on nonmetricity by only improving the statistics but keeping the same methodology. Within the general theoretical framework of the SME, complementary constraints on nonmetricity components may be feasible in the near future including limits on anisotropic components.

Acknowledgments

This work was supported in part by the DOE, by the NSF under grant No. PHY-1614545, by the IU Center for Spacetime Symmetries, by the IU Collaborative Research and Creative Activity Fund of the Office of the Vice President for Research, by the IU Collaborative Research Grants program, by the National Science Foundation of China under grant No. 11605056 and No. 11574082. RL acknowledges support from the Alexander von Humboldt Foundation. ZX is grateful for the hospitality of the IUCSS and the University of the Algarve.

References

[1] Abbott B P *et al.* 2016 *Phys. Rev. Lett.* **116** 061102
Abbott B P *et al.* 2016 *Phys. Rev. Lett.* **116** 241103
[2] Blagojević M and Hehl F W *eds.* 2013 *Gauge Theories of Gravitation* (London: Imperial College Press)
[3] Weyl H 1917 *Annalen der Physik IV* **364** [resp. **59**] 101–133
Weyl H 1918 Gravitation and Elektrizität *Sitzungsber. Preuss. Akad. Wiss.* 465–480
[4] Cartan E 1922 *C.R. Acad. Sci.* (Paris) **174** 593
[5] Tasson J D 2016 *Symmetry* **8** 111
[6] Colladay D and Kostelecký V A 1997 *Phys. Rev.* D **55** 6760
Colladay D and Kostelecký V A 1998 *Phys. Rev.* D **58** 116002
Kostelecký V A 2004 *Phys. Rev.* D **69** 105009
[7] Kostelecký V A and Russell N 2011 *Rev. Mod. Phys.* **83** 11
[8] Kostelecký V A, Russell N and Tasson J D 2008 *Phys. Rev. Lett.* **100** 111102
de Andrade L C G 2017 *Eur. Phys. J.* C **77** 401
[9] Lehnert R, Snow W M and Yan H 2014 *Phys. Lett.* B **730** 353 [Erratum: 2015 *Phys. Lett.* B **744** 415]
Yan H *et al.* 2015 *Phys. Rev. Lett.* **115** 182001
Ivanov A N and Snow W M 2017 *Phys. Lett.* B **764** 186
[10] Ni W-T 2010 *Rept. Prog. Phys.* **73** 056901
[11] Ramsey N F 1979 *Physica* (Amsterdam) **96A** 285
Vasilakis G, Brown J M , Kornack T W and Romalis M V 2009 *Phys. Rev. Lett.* **103** 261801
Jackson Kimball D F, Boyd A and Budker D 2010 *Phys. Rev.* A **82** 062714
Adelberger E G and Wagner T A 2013 *Phys. Rev.* D **88** 031101(R)
[12] Ni W-T 1979 *Phys. Rev.* D **19** 2260
[13] Neville D E 1980 *Phys. Rev.* D **21** 2075
Neville D E 1982 *Phys. Rev.* D **25** 573
Carroll S M and Field G B 1994 *Phys. Rev.* D **50** 3867
Raffelt G 2012 *Phys. Rev.* D **86** 015001
[14] Lämmerzahl C 1997 *Phys. Lett.* A **228** 223
[15] Mohanty S and Sarkar U 1998 *Phys. Lett.* B **433** 424
[16] Belyaev A S, Shapiro I L and do Vale M A B 2007 *Phys. Rev.* D **75** 034014
[17] Capozziello S, Cianci R, De Laurentis M and Vignolo S 2010 *Eur. Phys. J.* C **70** 341
[18] Mao Y, Tegmark M, Guth A H and Cabi S 2007 *Phys. Rev.* D **76** 104029
Flanagan E E and Rosenthal E 2007 *Phys. Rev.* D **75** 124016
Hehl F W, Obukhov Y N and Puetzfeld D 2013 *Phys. Lett.* A **377** 1775
[19] Canè F *et al.* 2004 *Phys. Rev. Lett.* **93** 230801
Heckel B R *et al.* 2006 *Phys. Rev. Lett.* **97** 021603
Heckel B R *et al.* 2008 *Phys. Rev.* D **78** 092006
[20] Foster J, Kostelecký V A and Xu R 2017 *Phys. Rev.* D **95** 084033
[21] Ne'eman Y and Hehl F W 1997 *Class. Quant. Grav.* **14** A251
Vitagliano V 2014 *Class. Quant. Grav.* **31** 045006
[22] Nico J S *et al.* 2005 *J. Res. NIST* **110** 137
[23] Kostelecký V A and Lehnert R 2001 *Phys. Rev.* D **63** 065008
Fittante A and Russell N 2012 *J. Phys.* G: *Nucl. Part. Phys.* **39** 125004
[24] Kostelecký V A and Mewes M 2013 *Phys. Rev.* D **88** 096006
[25] Lehnert R, Snow W M, Xiao Z and Xu R 2017 *Phys. Lett.* B **772** 865
[26] Foldy L L and Wouthuysen S A 1950 *Phys. Rev.* **78** 29
Kostelecký V A and Lane C D 1999 *J. Math. Phys.* **40** 6245
[27] Bass C D *et al.* 2009 *Nucl. Instrum. Meth.* A **612** 69
[28] Micherdzinska A M *et al.* 2011 *Nucl. Instrum. Meth.* A **631** 80
[29] Swanson H E and Schlamminger S 2010 *Meas. Sci. Technol.* **21** 115104

IOP Conf. Series: Journal of Physics: Conf. Series **952** (2018) 012014 doi:10.1088/1742-6596/952/1/012014

[30] Snow W M *et al.* 2011 *Phys. Rev.* C **83** 022501(R)
Snow W M *et al.* 2012 *Nuovo Cimento* C **35** 04
Snow W M *et al.* 2015 *Rev. Sci. Instrum.* **86** 055101
[31] Desplanques B 1998 *Phys. Rep.* **297** 1

IOP Conf. Series: Journal of Physics: Conf. Series **952** (2018) 012015 doi:10.1088/1742-6596/952/1/012015

Projectable Hořava Gravity in 2+1 Dimensions: Phases and Flows

Kevin T Grosvenor

Niels Bohr Institute, University of Copenhagen, Blegdamsvej 17, DK-2100 Copenhagen Ø, Denmark

E-mail: `kevin.grosvenor@nbi.ku.dk`

Abstract. We study the phase structure of projectable Hořava gravity in 2+1 dimensions and consider the implications of recent renormalization group calculations on the phase diagram.

1. Introduction

Next-generation experiments promise to test the assumptions of general relativity (GR) to ever greater precision putting ever tighter bounds on Lorentz-violating effects in gravity [1]. Above and beyond the theoretical aspirations of a potential renormalizable ultraviolet (UV) completion of GR, practicality dictates that we study nonrelativistic theories of quantum gravity in preparation for their collision with data. As a first step, in this work we study a simplified model of Hořava gravity.

Hořava gravity is a Lifshitz-type quantum field theory of gravity, in which time and space scale differently from one another [2]. Unlike the theory of General Relativity, Hořava's theory is power-counting renormalizable and is thus a candidate ultraviolet-complete theory of quantum gravity. Comparisons have also been drawn between Hořava gravity and the lattice approach of Causal Dynamical Triangulations (CDT) [3], which is a Monte Carlo method of simulating quantum gravity on a computer. As a point of comparison to CDT, one natural first observable to consider in Hořava gravity is the ground state, which, at the classical level, is described just by the geometry of the solutions to the equations of motion. We will examine the phases of these solutions as functions of the various parameters in the theory.

While a full-fledged calculation of the renormalization group flow of the theory in general remains a significant challenge, results have been obtained recently for the so-called projectable Hořava gravity theory in $2+1$ dimensions [4, 5]. We describe these results below and comment on their effect on the phase diagram.

2. Hořava Gravity

In Hořava gravity, spacetime is equipped with a global time coordinate t and a foliation structure by leaves of constant time. The natural geometric fields are the lapse function N, the shift vector N_i and the spatial metric g_{ij}. The theory is invariant under diffeomorphisms which preserve the foliation structure. The building blocks with which the action is constructed are the spatial Riemann tensor $R_{ijk\ell}$, the acceleration vector a_i and the extrinsic curvature tensor K_{ij}. The latter two are defined as

$$a_i = \frac{\nabla_i N}{N}, \tag{1}$$

$$K_{ij} = \frac{1}{2N}(\partial_t g_{ij} - \nabla_i N_j - \nabla_j N_i), \tag{2}$$

where ∇_i is the covariant derivative with respect to g_{ij} and Latin indices are raised and lowered by the spatial metric. Here, we will immediately specialize to the projectable case, in which $N(t, \mathbf{x}) = N(t)$ or $a_i = 0$. In this case, the action is given by

$$S = \frac{1}{2\kappa^2} \int dt\, d^D\mathbf{x}\, N\sqrt{g}\Big(K_{ij}K^{ij} - \lambda K^2 - V(g_{ij}, \nabla_i)\Big), \tag{3}$$

where D is the space dimension and V is a potential function of the metric and connection.

In the vicinity of a Gaussian fixed point, the theory gains an anisotropic scaling symmetry with the dynamical critical exponent z, whereby time and space scale as

$$\begin{aligned} t &\to b^z t, \\ \mathbf{x} &\to b\mathbf{x}. \end{aligned} \tag{4}$$

In the ultraviolet (UV), z is classically the maximum number of pairs of derivatives in a term in V (assuming parity invariance). With classical dimensions adapted to this scaling, terms in V with greater than, equal to, and less than D pairs of derivatives are irrelevant, marginal, and relevant, respectively.

Firstly, in $D = 2$, the potential can only be a function of the Ricci scalar R since the Riemann tensor is is in fact proportional to R. Secondly, considering only marginal and relevant terms,

$$V_{D=2} = \frac{\alpha}{2}R^2 - \beta R + 2\gamma. \tag{5}$$

In $D = 3$, the Riemann tensor depends on both the Ricci scalar and the Ricci tensor. In addition to the terms above, there are three marginal terms of R^3 type and two of $(\nabla R)^2$ type, and one additional relevant term $R_{ij}R^{ij}$. To avoid this proliferation of terms, we will consider only projectable Hořava gravity in $2+1$ dimensions given by (3) and (5).

Theories like this, which contain multiple terms quadratic in fields but with different powers of spatial derivative, naturally give rise to multiple types of observers, which measure different scaling dimensions. In our case, the UV observer sets $\alpha = 1$, whereas the IR observer sets $\beta = 1$. The UV observer sees $z = 2$ scaling between time and space, whereas the IR observer sees $z = 1$ scaling and effectively sets $\kappa^2 = 8\pi G_N$, where G_N is Newton's constant, so that the action resembles GR. Note that full diffeomorphism invariance would further require $\lambda = 1$.

3. The Phases

We take the following Friedmann-Lemaître-Robertson-Walker (FLRW) ansatz for the metric,

$$\begin{aligned} N &= 1, \\ N_i &= 0, \\ g_{ij} &= f(t)\hat{g}_{ij}, \end{aligned} \tag{6}$$

where $\hat{g}_{ij}$ is a constant curvature metric with

$$\hat{R}_{ij} = k\hat{g}_{ij}, \tag{7}$$

IOP Conf. Series: Journal of Physics: Conf. Series **952** (2018) 012015 doi:10.1088/1742-6596/952/1/012015

and k can be normalized to $k = \pm 1, 0$.

Evaluated on the FLRW ansatz, the Friedmann equation (the N equation of motion) is

$$\kappa^2 f^2 \mathcal{H} = -\varepsilon \dot{f}^2 + \alpha - \beta k f + \gamma f^2 = 0, \tag{8}$$

where $\mathcal{H}$ is the Hamiltonian density and

$$\varepsilon \equiv \frac{2\lambda - 1}{4}. \tag{9}$$

The g_{ij} equation of motion reduces to $\dot{\mathcal{H}}/\dot{f} = 0$, or simply $\dot{\mathcal{H}} = 0$, so long as f is not constant.

Before solving these equations, let us first determine an appropriate set of parameters with which to write the solution. Note that α and κ^2 do not have an independent meaning and only the combination

$$\mathcal{G} \equiv \sqrt{\frac{2\kappa^4}{\alpha}}, \tag{10}$$

matters, where $\mathcal{G}$ is the notation used in [5]. Even if we measure time dimensions T and space dimensions L separately, $\mathcal{G}$ remains dimensionless. The other key dimensionless parameter is ε. Similarly, the appropriate quantity parametrizing the Einstein-Hilbert term in the action is

$$\eta \equiv \frac{\beta}{2\kappa^2}, \tag{11}$$

which has dimensions of $[\eta] = T^{-1}$. A convenient measure of the cosmological constant is

$$\Omega \equiv \frac{2\gamma}{\mathcal{G}^2} = \frac{\alpha\gamma}{\kappa^4}, \tag{12}$$

which has dimensions of $[\Omega] = T^{-2}$. It also turns out to be convenient to define

$$\begin{aligned} \tau &\equiv \frac{\mathcal{G}}{\sqrt{2\varepsilon}} t, \\ \zeta(\tau) &\equiv \frac{\mathcal{G}^2}{2\kappa^2} f(t). \end{aligned} \tag{13}$$

With this definition, $\zeta(\tau)$ has dimension $[\zeta] = T$.

Firstly, when $\varepsilon = 0$, the solution is constant

$$\zeta(\tau)|_{\varepsilon=0} = \frac{k}{\eta}, \tag{14}$$

as long as this is positive and $\eta^2 = \Omega$. Henceforth, we consider $\varepsilon \neq 0$.

3.1. Real Time

Real time corresponds to $\mathcal{G}^2$ and ε having the same sign. We assume that they are both positive since the case when they are both negative can easily be retrieved by analytic continuation.

Consider the three regions:

$$\begin{aligned} &\text{Region 1: } \Omega < 0, \\ &\text{Region 2: } \eta^2 > \Omega > 0, \\ &\text{Region 3: } \Omega > \eta^2 > 0. \end{aligned} \tag{15}$$

In region 1, the solution for $\zeta(\tau)$ reads

$$\zeta_1(\tau) = \frac{1}{|\Omega|}\left[-\eta k + |k|\sqrt{\eta^2 + |\Omega|}\,\cos\left(\sqrt{|\Omega|}\,\tau\right)\right]. \tag{16}$$

In region 2, there are two potential solutions

$$\zeta_2(\tau) = \frac{1}{\Omega}\left[\eta k \pm |k|\sqrt{\eta^2 - \Omega}\,\cosh\left(\sqrt{\Omega}\,\tau\right)\right]. \tag{17}$$

If $\eta k > 0$, then the + solution is an eternal universe that is infinite flat space in the infinite past, curves into a sphere ($k > 0$) or hyperbolic plane ($k < 0$) of decreasing size up to a minimum at $\tau = 0$, and then grows infinitely again towards the future. In contrast, the − solution is the direct analytic continuation of the big bang/big crunch solution (16). On the other hand, if $\eta k < 0$, the − solution is not permissible since it is always negative, and the + solution consists of two disconnected pieces: a big crunch, then a big bang.

In region 3, two solutions exist: one is an ever-growing big bang and the other is the time-reversed version (big crunch),

$$\zeta_3(\tau) = \frac{1}{\Omega}\left[\eta k \pm |k|\sqrt{\Omega - \eta^2}\,\sinh\left(\sqrt{\Omega}\,\tau\right)\right]. \tag{18}$$

3.2. Imaginary Time

Imaginary time corresponds to $\mathcal{G}^2$ and ε having opposite sign. Suppose $\mathcal{G}^2 > 0$ and $\varepsilon < 0$ (the other choice is related by analytic continuation).

Writing $\tau = i\tau_E$, the solution in region 1 is the same as (16) but with τ replaced with τ_E and cos replaced with cosh. This is similar to the eternal universe solution in region 2 in the real time case (17). The solution in region 2 is the same as (17) but with τ again replaced with τ_E and with a cos instead of cosh. The two signs ± are actually identical now and there is only one solution, which is oscillating forever between a maximum and minimum radius. Additionally, only spherical solutions ($k > 0$) are allowed when $\eta > 0$ and only hyperbolic solutions ($k < 0$) are allowed when $\eta < 0$. Finally, and perhaps most interestingly, no solution exists in region 3. In this case, we call region 3 the "Forbidden Region".

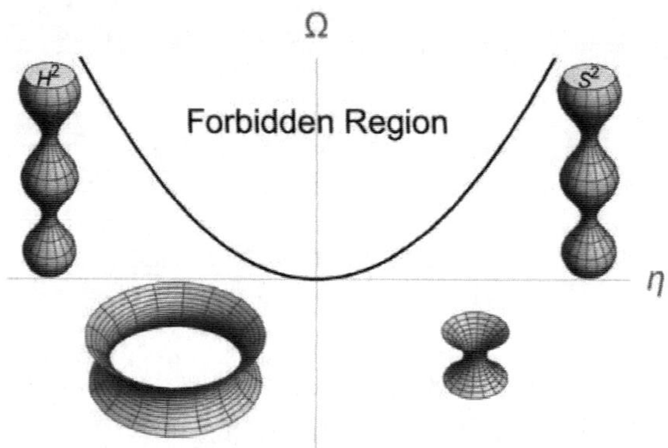

Figure 1. Imaginary time phase diagram. Region 1 ($\Omega < 0$) is flat space at $\tau = \pm\infty$ and curves to a minimum radius at $\tau = 0$. Region 2 ($\eta^2 > \Omega > 0$) is oscillating spheres for $\eta > 0$ and hyperbolic disks for $\eta < 0$. No solution of the form (6) exists in the Forbidden Region.

IOP Conf. Series: Journal of Physics: Conf. Series **952** (2018) 012015 doi:10.1088/1742-6596/952/1/012015

4. The Flows

The renormalization group flow within the subspace $(\mathcal{G}, \lambda)$ of dimensionless couplings was computed in [5] with the results

$$\begin{aligned} \beta_{\mathcal{G}} &= -\frac{16 - 33\lambda + 18\lambda^2}{64\pi(1-\lambda)^2}\sqrt{\frac{1-\lambda}{1-2\lambda}}\,\mathcal{G}^2, \\ \beta_\lambda &= \frac{15 - 14\lambda}{64\pi}\sqrt{\frac{1-2\lambda}{1-\lambda}}\,\mathcal{G}. \end{aligned} \tag{19}$$

There is an asymptotically free ($\mathcal{G} = 0$) UV fixed at $\lambda = \frac{15}{14}$. For $1 < \lambda < \frac{15}{14}$, the theory flows towards strong coupling at the GR value of $\lambda = 1$. For $\lambda > \frac{15}{14}$, the theory flows towards ever increasing λ and $\mathcal{G}$. There is also an interesting family of UV fixed points at the value $\lambda = \frac{1}{2}$ parametrized by the value of the coupling $\tilde{\mathcal{G}} = \mathcal{G}/\sqrt{1-2\lambda}$.

At each point on this two-dimensional RG flow diagram sits a two-dimensional phase diagram in the dimensionful couplings (η, Ω). The effect of the values of $\mathcal{G}$ and λ are parametrized in the definitions of τ and $\zeta(\tau)$ in (13). At the mean-field level, the RG flows in the (η, Ω) plane are controlled by the dimensions of η and Ω and, since $[\Omega] = T^{-2} = [\eta^2]$, the flow lines are simply parabolas from the origin. Taking one-loop effects into account, it was shown in [4] that η does not flow, but Ω gains an anomalous dimension (measured in units of momentum)

$$\delta_\Omega = \frac{1}{32\pi}\left(\frac{1-2\lambda}{1-\lambda}\right)^{3/2}\mathcal{G}. \tag{20}$$

Therefore, instead of parabolas $\Omega \propto \eta^2$, the flow lines are altered to

$$\Omega \propto \eta^{2-\frac{\delta_\Omega}{2}}. \tag{21}$$

5. Discussion

There is a striking similarity between the phase diagram in imaginary time of projectable Hořava gravity in 3 dimensions and that of CDT in 4 dimensions (after Wick rotation). The eternal universe in region 1 corresponds to the de Sitter-like phase (phase C) of CDT. The oscillating phase in region 2 with $\eta > 0$ corresponds to the oscillating phase (phase A) of CDT.

The forbidden region would correspond to the "pancake" phase (phase B) of CDT in which the entire universe lives inside one time step. Thus, one may conjecture that the forbidden region represents a "topological phase of lapse" in which $N \to 0$ and the spatial metric is static. Indeed, when measured with respect to any finite lapse, the history of this static universe collapses to zero time. In this case, the Hamiltonian is no longer forced to vanish and should, in principle, be minimized in the ground state. Why the Hamiltonian should be deconfined away from 0 as one crosses the phase transition line into the forbidden region is an outstanding mystery.

The identification of a fourth phase (the so-called "bifurcation" phase or phase D) in CDT between phases B and C naturally leads one to identify this with the oscillating hyperbolic phase in region 2 with $\eta < 0$. Of course, CDT cannot produce non-compact universes, but it instead seems to observe a metric signature change [6].

This comparison begs the question of the difference in spacetime dimension between our analysis herein and the simulations of CDT. If CDT were indeed simulating Hořava gravity in $3+1$ dimensions, either a zoo of phases have eluded observation or CDT is seeing signs of a fixed point at which many of the couplings vanish. To examine this possibility further, we would need the RG flow of the full theory in $3+1$ dimensions, which is hopefully forthcoming.

IOP Conf. Series: Journal of Physics: Conf. Series **952** (2018) 012015 doi:10.1088/1742-6596/952/1/012015

Acknowledgements

It is a pleasure to thank Charles M. Melby-Thompson and Petr Hořava for their collaboration on this project. I would also like to thank the University of Algarve, CENTRA and organizers of the International Workshop on CPT and Lorentz Symmetry in Field Theory at Faro, Portugal. The work of K.T.G. is supported by the FNU grant number DFF-6108-00340.

References

[1] Williams J, Chiow S, Mueller H and Yu N 2016 *New J. Phys.* **18** 025018
[2] Hořava P 2009 *J. High Energy Phys.* JHEP03(2009)020
[3] Ambjørn J, Görlich A, Jordan S, Jurkiewicz J and Loll R 2010 *Phys. Lett.* B **690** 413-9
[4] Griffin T, Grosvenor K T, Melby-Thompson C M and Yan Z 2017 *J. High Energy Phys.* JHEP06(2017)004
[5] Barvinsky A O, Blas D, Herrero-Valea M, Sibiryakov S M and Steinwachs C F 2016 *Phys. Rev.* D **93** 064022
[6] Ambjørn J, Coumbe D N, Gizbert-Studnicki J and Jurkiewicz J 2015 *J. High Energy Phys.* JHEP08(2015)033

Quantum information aspects of noncommutative quantum mechanics

Orfeu Bertolami, Alex E Bernardini and Pedro Leal

Departamento de Física e Astronomia, Faculdade de Ciências da Universidade do Porto, Rua do Campo Alegre, 687,4169-007 Porto, Portugal

E-mail: `orfeu.bertolami@fc.up.pt`

Abstract. Some fundamental aspects related with the construction of Robertson-Schrödinger-like uncertainty-principle inequalities are reported in order to provide an overall description of quantumness, separability and nonlocality of quantum systems in the noncommutative phase-space. Some consequences of the deformed noncommutative algebra are also considered in physical systems of interest.

1. Introduction

In searching for a broader understanding of the transition from quantum to classical description, noncommutative (NC) algebra deformations of the standard quantum mechanics (QM) plays a relevant role. The NC extension of QM is characterized by the deformation of the Heisenberg-Weyl algebra,

$$[\hat{x}_i, \hat{x}_j] = i\theta_{ij}, \quad [\hat{x}_i, \hat{p}_j] = i\hbar\delta_{ij}, \quad [\hat{p}_i, \hat{p}_j] = i\eta_{ij}, \quad i = 1, ...n, \tag{1}$$

where θ_{ij} and η_{ij} are skew-symmetric matrices with real entries, whereas in the standard quantum mechanics one has

$$[\hat{q}_i, \hat{q}_j] = 0, \quad [\hat{q}_i, \hat{k}_j] = i\hbar\delta_{ij}, \quad [\hat{k}_i, \hat{k}_j] = 0, \quad i = 1, ...n, \tag{2}$$

with the two algebras being connected by the so-called Seiberg-Witten (SW) map [1],

$$\hat{x}_i = \nu\hat{q}_i - \frac{\theta}{2\nu\hbar}\epsilon_{ij}\hat{k}_j, \quad \hat{p}_i = \mu\hat{k}_i + \frac{\eta}{2\mu\hbar}\epsilon_{ij}\hat{q}_j, \tag{3}$$

which allows, through constants ν and μ, mapping NC effects in terms of standard QM variables.

From the perspective of a deformed Heisenberg-Weyl algebra of QM [2, 3, 4], several issues on NC QM related to missing information in gaussian quantum systems [5, 6, 7], quantum correlations and information collapse [8, 9, 10], and violations of the uncertainty relations [11, 12, 13, 14] have been investigated.

The theoretical framework of the NC QM [2, 3, 15, 17, 18] is suitably formulated in the Weyl-Wigner-Groenewold-Moyal (WWGM) formalism for QM [19, 20, 21], in a theory supported by a $2n$-dim phase space where coordinate and momentum variables obey a NC algebra. Through the WWGM formalism, the NC QM has been used to understand the coupling and decoherence aspects of 2-dim quantum oscillators [4, 8], the quantum Hall effect [22] and the Landau level problem [23], corrections to the gravitational quantum well [15] and nuclear matter [16]. Furthermore, besides its relevance in the context of quantum gravity and string theory [1, 24, 25], NC extensions of quantum cosmology have also been considered in the investigation of the black-hole singularity [11, 26, 27], as well as of gauge invariance [28, 29] and the equivalence principle [29, 30].

2. Quantumness

Considering a bipartite quantum system described in terms of a $2n_A$-dimensional subsystem A and a $2n_B$-dimensional subsystem B, with $n_A + n_B = n$, one can write the collective degrees of freedom of the composite quantum system as $\widehat{z} = (\widehat{z}^A, \widehat{z}^B)$, where $\widehat{z}^{A,B} = (\widehat{x}_1^{A,B}, \cdots, \widehat{x}_{n_{A,B}}^{A,B}, \widehat{p}_1^{A,B}, \cdots, \widehat{p}_{n_{A,B}}^{A,B})$ correspond to the generalized coordinates of the two subsystems [5]. The NC phase space algebra given by

$$[\widehat{z}_i, \widehat{z}_j] = i\,\Omega_{ij}, \qquad i, j = 1, \cdots, 2n, \tag{4}$$

is described by the matrix, $\mathbf{\Omega} = [\Omega_{ij}] \equiv \mathrm{Diag}\left[\mathbf{\Omega^A},\, \mathbf{\Omega^B}\right]$, where [2, 5]

$$\mathbf{\Omega^{A,B}} = \begin{pmatrix} \mathbf{\Theta^{A,B}} & \mathbf{I^{A,B}} \\ -\mathbf{I^{A,B}} & \mathbf{\Pi^{A,B}} \end{pmatrix}, \tag{5}$$

are real skew-symmetric non-singular $2n_{A,B} \times 2n_{A,B}$ matrices with $\mathbf{\Theta^{A,B}} = \left[\theta_{ij}^{A,B}\right]$ and $\mathbf{\Pi^{A,B}} = \left[\eta_{ij}^{A,B}\right]$ measuring the noncommutativity of the position and momentum sectors of the subsystems A and B, respectively[1]. The generalization of the NC structure from Eqs. (1)-(3) is indeed formulated in terms of commuting variables through the linear Darboux transformation (DT) (also referred to as SW map), Eq. (3), $\widehat{z} = \mathbf{S}\widehat{\zeta}$, with $\mathbf{S} \in \mathrm{GL}(2n)$ such that $\mathbf{S} = \mathrm{Diag}\left[\mathbf{S^A},\, \mathbf{S^B}\right]$, where $\widehat{\zeta} = (\widehat{\zeta}^A, \widehat{\zeta}^B)$, with $\widehat{\zeta}^{A,B} = (\widehat{q}_1^{A,B}, \cdots, \widehat{q}_{n_{A,B}}^{A,B}, \widehat{k}_1^{A,B}, \cdots, \widehat{k}_{n_{A,B}}^{A,B})$ satisfying the usual QM commutation relations:

$$\left[\widehat{\zeta}_i, \widehat{\zeta}_j\right] = iJ_{ij}, \qquad i, j = 1, \cdots, 2n, \tag{6}$$

where $\mathbf{J} = [J_{ij}] = \mathrm{Diag}\left[\mathbf{J^A}, \mathbf{J^B}\right]$ with

$$\mathbf{J^{A,B}} = -(\mathbf{J^{A,B}})^T = -(\mathbf{J^{A,B}})^{-1} = \begin{pmatrix} \mathbf{0^{A,B}} & \mathbf{I^{A,B}} \\ -\mathbf{I^{A,B}} & \mathbf{0^{A,B}} \end{pmatrix}, \tag{7}$$

which are $2n_{A,B} \times 2n_{A,B}$ standard symplectic matrices. Like the SW map, also the map $\mathbf{S}$ is not uniquely defined: if one composes $\mathbf{S}$ with block-diagonal canonical transformations, an equivalent DT is obtained, and one recovers the correspondence given by $\mathbf{\Omega} = \mathbf{SJS}^T$ [2, 5].

For a composite system described by the density matrix ρ, function of the NC variables $\widehat{z}$, the DT yields the density matrix $\widetilde{\rho}(\widehat{\zeta}) = \rho\left(\mathbf{S}\widehat{\zeta}\right)$, which is associated with the Wigner function $W\widetilde{\rho}(\zeta)$ such that, upon an inversion of the DT, one obtains the NC Wigner function [2]:

$$W\rho(z) = \frac{1}{\sqrt{\det \mathbf{\Omega}}} W\widetilde{\rho}(\mathbf{S}^{-1}z) \quad \leftrightarrow \quad W^{NC}(z) = \frac{1}{\sqrt{\det \mathbf{\Omega}}} W(\mathbf{S}^{-1}z), \tag{8}$$

[1] Here $\mathbf{I^{A,B}}$ is the $n_{A,B} \times n_{A,B}$ identity matrix and we have set $\hbar = 1$.

IOP Conf. Series: Journal of Physics: Conf. Series **952** (2018) 012016 doi:10.1088/1742-6596/952/1/012016

obtained by *turning on* the NC algebra [5, 6, 11]. In the simplified scenario where Wigner functions are identified with gaussian functions, if $\mathbf{\Sigma}$ denotes the covariance matrix of $W\rho$ and $\widetilde{\mathbf{\Sigma}}$ that of $W\widetilde{\rho}$, then the two are related by:

$$\mathbf{\Sigma} = \mathbf{S}\widetilde{\mathbf{\Sigma}}\mathbf{S}^T. \tag{9}$$

A necessary condition for the phase-space Wigner function $W\widetilde{\rho}$ with covariance matrix $\widetilde{\mathbf{\Sigma}}$ to be quantum mechanically admissible is that it satisfies the Robertson-Schrödinger uncertainty principle (RSUP) [12, 13],

$$\widetilde{\mathbf{\Sigma}} + (i/2)\mathbf{J} \geq 0, \tag{10}$$

that is, in the representation of a $2n \times 2n$ positive matrix in C^{2n}. For $W\rho$ to be an equally admissible NC Wigner function, from Eqs. (9) and (10), one concludes that it has to satisfy the NC RSUP,

$$\mathbf{\Sigma} + (i/2)\mathbf{\Omega} \geq 0, \tag{11}$$

which, for NC gaussian states, is also a sufficient condition for defining the quantum behavior. Here it is worth to mention that a matrix version of the Ozawa uncertainty principle [12] can also accomodate any NC structure in the phase space [13]. In particular, it leads to NC corrections to lowest order for two measurement interactions: the backaction evading quadrature amplifier and noiseless quadrature transducers [14].

3. Entanglement

Using similar tools [5], gaussian separability is identified through a manipulation supported by the introduction of a $2n \times 2n$ matrix $\mathbf{\Lambda} = \mathrm{Diag}\left[\mathbf{I^A},\, \mathbf{\Lambda^B}\right]$, with $\mathbf{\Lambda^B} = \mathrm{Diag}\left[\mathbf{I},\, -\mathbf{I}\right]$. The transformation $\zeta \mapsto \mathbf{\Lambda}\zeta$ amounts to a mirror reflection of B momenta and, through the PPT criterion [5], if a Wigner function $W\widetilde{\rho}(\zeta)$ is the one of a separable state, then the transformation

$$W\widetilde{\rho}(\zeta) \mapsto W\widetilde{\rho}'(\zeta) = W\widetilde{\rho}(\mathbf{\Lambda}\zeta), \tag{12}$$

leads to an equivalent Wigner function. If the state $\widetilde{\rho}$ is separable then

$$\widetilde{\mathbf{\Sigma}}' + (i/2)\mathbf{J} \geq 0, \tag{13}$$

where $\widetilde{\mathbf{\Sigma}'}$ is the covariance matrix of $W\widetilde{\rho}'(\zeta)$. In terms of the NC variables one has

$$W\rho(z) \mapsto W\rho'(z) = W\rho(\mathbf{D}z), \quad \text{with} \quad W\rho'(z) = \frac{1}{\sqrt{\det \mathbf{\Omega}}} W\widetilde{\rho}'(\mathbf{S}^{-1}z), \tag{14}$$

where $\mathbf{D} = \mathbf{D}^{-1} = \mathbf{S\Lambda S}^{-1} = \mathrm{Diag}[\mathbf{I^A},\, \mathbf{S^B\Lambda^B(S^B)}^{-1}]$, and the covariance matrices $\mathbf{\Sigma}'$ and $\widetilde{\mathbf{\Sigma}}'$ of $W\rho'$ and $W\widetilde{\rho}'$ are related by $\mathbf{\Sigma}' = \mathbf{S}\widetilde{\mathbf{\Sigma}}'\mathbf{S}^T$, which results into a separability condition given exclusively in terms of the NC objects,

$$\mathbf{\Sigma}' + (i/2)\mathbf{\Omega} \geq 0\ . \tag{15}$$

By noticing that $\mathbf{\Sigma}' = \mathbf{D\Sigma D}^T$ and defining $\mathbf{\Omega}' = \mathbf{D}^{-1}\mathbf{\Omega}(\mathbf{D}^T)^{-1}$ one also has

$$\mathbf{\Sigma} + (i/2)\mathbf{\Omega}' \geq 0\ . \tag{16}$$

with $\mathbf{\Omega}'$ written as

$$\mathbf{\Omega}' = \mathrm{Diag}\left[\mathbf{\Omega^A},\, -\mathbf{\Omega^B}\right], \tag{17}$$

where one has used the definition of $\mathbf{\Omega}$ and the fact that $\mathbf{\Lambda^B}$ is an anti-symplectic transformation, i. e. $\mathbf{\Lambda^B J^B \Lambda^B} = -\mathbf{J^B}$. By investigating the properties of W and W^{NC}, one concludes that a state that is separable in usual QM might be entangled in a NC extension. For gaussian functions, this is a purely kinematical effect, related to the structure of the NC algebra.

4. Bell Inequalities

The above quantum description can be reduced to a bipartite quantum system described in terms of a modes A and B, with collective degrees of freedom $\widehat{z} = (\widehat{z}^A, \widehat{z}^B)$, where $\widehat{z}^A = (\widehat{x}, \widehat{p}_x)$ and $\widehat{z}^B = (\widehat{y}, \widehat{p}_y)$, which do help to identify the influence that noncommutativity may have on nonlocal features. In order to quantify the non-locality of a state through the Clauser, Horne, Shimony and Holt (CHSH) inequality [31], one must have an explicit definition for the Bell operator. The Bell operator is a functional of the Wigner function which parameterizes a linear combination of four expectation values [31],

$$\mathcal{B} = \frac{\pi^2}{4} \left[W(0,0) + W(\alpha_1, 0) + W(0, \alpha_2) - W(\alpha_1, \alpha_2) \right] , \tag{18}$$

where $\alpha_{1,2}$ are the phase-space complex amplitudes associated with the operators $\hat{\alpha}_1 = \hat{x} + i\hat{p}_x$ and $\hat{\alpha}_2 = \hat{y} + i\hat{p}_y$, that carry two degrees of freedom, $\hat{\alpha}$ and $\hat{\alpha}^*$ for each mode, 1 and 2, and $W(\alpha_1, \alpha_2)$ is the commutative Wigner function of the state calculated in (α_1, α_2). Local theories admit a description in terms of local hidden variables identified by α_1 and α_2 (in Eq.(18)). Straightforward calculations [31] allow to show the dichotomic behaviour of the Bell operator corresponds to the boundary value for non-locality given by $|\mathcal{B}_{min}| > 2$.

In order to implement the NC effects on the Bell functional, one assumes that the Wigner function in Eq. (18) belongs to the set of NC Wigner functions $\mathcal{W}^{NC}$ [32]. The same relation is valid for all observables $\mathcal{O}$ that are completely symmetric in the NC variables[2]. Then the Weyl symbol $\mathcal{O}^{NC}(z)$ of $\widehat{\mathcal{O}}^{NC}$ and the Weyl symbol $\mathcal{O}(\xi)$ of $\widehat{\mathcal{O}}$ are given by the same function, i.e. $\mathcal{O}^{NC} = \mathcal{O}$, and so the average value functionals of the two observables are the same [32]. That is, given a *commutative* operator $\widehat{\mathcal{O}} = \mathcal{O}(\widehat{\xi})$, its NC version is defined by $\widehat{\mathcal{O}}^{NC} = (\mathcal{O}(\widehat{z}))_+$, and thus the expectation values of both operators are given by the same functional of the Wigner function in the NC form.

The time evolution of the NC average values $\langle . \rangle$ is determined from the dynamics of the NC Wigner function given by

$$\dot{W}^{NC}(z;\, t) = [H^{NC}(z), W^{NC}(z;\, t)]_{NC} ,$$

where $H^{NC}(z) = \tilde{H}(\mathbf{S}^{-1}z)$, $\tilde{H}(\xi)$ is the Weyl symbol of $H(\mathbf{S}\widehat{\xi})$, and the (commutative and the NC) Moyal brackets shall determine two different dynamics for the Wigner functions. To quantifiy the impact of noncommutativity on the dynamics of the Bell function, one assumes that at the initial time the commutative and the NC configurations are the same. Therefore, at $t = 0$ one has $W(x,\, p;\, t = 0) = W^{NC}(x,\, p;\, t = 0)$, and so $\mathcal{B}^{NC}(t = 0) \equiv \mathcal{B}(W^{NC}(x,\, p;\, t = 0)) = \mathcal{B}(W(x,\, p;\, t = 0)) \equiv \mathcal{B}^C(t = 0)$. In this case, the measurement of expectation values of the fundamental variables is not influenced by the *a priori* knowledge of the NC QM structure. The only restriction is that the (NC) Wigner function yields the observed averaged values, and the time evolution of commutative and NC Wigner functions can be obtained in order to compute $\mathcal{B}^{NC}(t) = \mathcal{B}(W^{NC}(x,\, p;\, t))$, which can be compared to $\mathcal{B}^C(t) = \mathcal{B}(W(x,\, p;\, t))$. However, due to the non-trivial nature of the Bell operator, the inclusion of NC effects into its properties deserves further investigations. The results obtained in Ref. [32] indicate that the non-locality encountered in the NC case is qualitatively similar to the one of the usual QM.

5. Additional Issues and Outlook

Extending the known results of QM to the phase space NC QM framework requires investigating various quantum-mechanical phenomena in order to fully understand the implications of the

[2] Let $\widehat{\mathcal{O}}^{NC} = \mathcal{O}(\widehat{z})$ satisfy $\widehat{\mathcal{O}}^{NC} = (\widehat{\mathcal{O}}^{NC})_+$, where the subscript + denotes symmetrization with respect to the NC variables, and let $\widehat{\mathcal{O}} = \mathcal{O}(\widehat{\xi})$.

algebra deformation. For instance, examining whether gauge invariance or the equivalence principle hold under the deformation of the Heisenberg-Weyl algebra is particularly relevant. Regarding the former, when considering a particle in a magnetic field, it has been shown that in order to preserve gauge invariance, one must impose the vanishing of the configuration space noncommutative parameter, θ [29]. This follows from the fact that the Hamiltonian for the problem,

$$\hat{H} = \frac{1}{2m}\left[\boldsymbol{p} - q\boldsymbol{A}(\boldsymbol{x})\right]^2, \tag{19}$$

under a DT gives rise to terms that are not gauge invariant. As for the equivalence principle, it is shown that, when a gravitational field is considered in the single particle Hamiltonian, a suitable change to an accelerated frame of reference, $x \to x' = x + \sigma(t)$, is able to absorb this potential with only a phase difference on the wave function, which consequently is negligible for physical observables. Thus, the weak equivalence principle is recovered in NCQM [29].

Other known results of standard QM should be addressed in the context of NCQM. Interesting cases involve issues related with the measurement problem [33]. This will be pursued in a future work.

Acknowledgments

The work of OB is partially supported by the COST action MP1405. The work of AEB is supported by the Brazilian agency FAPESP (grant 17/02294-2). The work of PL is supported by FCT (Portuguese Fundação para Ciência e Tecnologia) grant PD/BD/135005/2017.

References

[1] Seiberg N and Witten E 1999 *J. High Energy Phys.* JHEP09(1999)032
[2] Bastos C, Bertolami O, Dias N C and Prata J N 2008 *J. Math. Phys.* **49** 072101
[3] Gamboa J, Loewe M and Rojas J C 2001 *Phys. Rev.* D **64** 067901
[4] Rosenbaum M and David Vergara J 2006 *Gen. Rel. Grav.* **38** 607
[5] Bastos C, Bernardini A E, Bertolami O, Dias N C and Prata J N 2013 *Phys. Rev.* D **88** 085013
[6] Bastos C, Bernardini A E, Bertolami O, Dias N C and Prata J N 2016 *Phys. Rev.* D **93** 104055
[7] Bernardini A E and Mizrahi S S2015 *Physica Scripta* **90**, 074011
[8] Bernardini A E and Bertolami O 2013 *Phys. Rev.* A **88** 012101
[9] Santos J F G, Bernardini A E and Bastos C 2015 *Physica* **A438** 340
[10] Santos J F G and Bernardini A E 2016 *Physica* **A 445** 75
[11] Bastos C, Bertolami O, Dias N C and Prata J N 2008 *Phys. Rev.* D **78** 023516
[12] Bastos C, Bernardini A E, Bertolami O, Dias N C and Prata J N 2014 *Phys. Rev.* A **89** 042112
[13] Bastos C, Bernardini A E, Bertolami O, Dias N C and Prata J N 2014 *Phys. Rev.* D **90** 045023
[14] Bastos C, Bernardini A E, Bertolami O, Dias N C and Prata J N 2015 *Phys. Rev.* D **91** 065036
[15] Bertolami O, Rosa J G, de Aragão C M L, Castorina P and Zappalà D 2005 *Phys. Rev.* D **72** 025010
Bertolami O, Rosa J G, de Aragão C M L, Castorina P and Zappalà D 2006 *Mod. Phys. Lett.* A **21** 795
[16] Bertolami O and Mariji H 2015 *Int. J. Mod. Phys.* **A** 30 1550191
[17] Dias N C and Prata J N 2009 *Annals Phys.* **324** 73
[18] Bastos C and Bertolami O 2008 *Phys. Lett.* A **372** 5556
[19] Groenewold H 1946 *Physica* **12** 405
[20] Moyal J 1949 *Proc. Camb. Phil. Soc.* **45** 99
[21] Wigner E 1932 *Phys. Rev.* **40** 749
[22] Prange R and Girvin S 1987 *The Quantum Hall Effect* (New York: Springer)
[23] Douglas M R and Nekrasov N A 2001 *Rev. Mod. Phys.* **73** 977
[24] Connes A, Douglas M R and Schwarz A 1998 *J. High Energy Phys.* JHEP02(1998)003
[25] Douglas M R and Hull C 1998 *J. High Energy Phys.* **02** JHEP02(1998)008
[26] Bastos C, Bertolami O, Dias N C and Prata J N 2009 *Phys. Rev.* D **80** 124038
[27] Leal P, Bertolami O and Bernardini A E 2017 (*Preprint* 1710.01172)
[28] Bertolami O, Queiroz R 2011 *Phys. Lett.* A **4116**
[29] Bertolami O, Leal P 2015 *Phys. Lett.* B **750** 6-11
[30] Bastos C, Bertolami O, Dias N C and Prata J N 2011 *Class. Quant. Grav.* **28** 125007
[31] Clauser H F, Horne M A, Shimony A and Holt R A 1969 *Phys. Rev. Lett.* **23** 880

[32] Bastos C, Bernardini A E, Bertolami O, Dias N C and Prata J N 2016 *Phys. Rev.* D **93** 104055
[33] Ballentine L E 1998 *Quantum Mechanics: A modern development* (World Scientific: Singapore)

IOP Conf. Series: Journal of Physics: Conf. Series **952** (2018) 012017 doi:10.1088/1742-6596/952/1/012017

Lorentz symmetry violation with higher-order operators and renormalization

J R Nascimento[1], A Yu Petrov[1] and C M Reyes[2]

[1] Departamento de Física, Universidade Federal da Paraíba Caixa Postal 5008, 58051-970, João Pessoa, Paraíba, Brazil
[2] Departamento de Ciencias Básicas, Universidad del Bío Bío, Casilla 447, Chillán, Chile

E-mail: `creyes@ubiobio.cl`

Abstract. Effective field theory has shown to be a powerful method in searching for quantum gravity effects and in particular for CPT and Lorentz symmetry violation. In this work we study an effective field theory with higher-order Lorentz violation, specifically we consider a modified model with scalars and modified fermions interacting via the Yukawa coupling. We study its renormalization properties, that is, its radiative corrections and renormalization conditions in the light of the requirements of having a finite and unitary S-matrix.

1. Introduction

Quantum gravity effects taking place at experiments at the energy scale m are expected to be suppressed by the Planck mass $m_P = 10^{19}$ GeV with the factor m/m_P. Such violent suppression leads us to consider new strategies in order to be able to detect possible signals from the Planck scale. The strategy encoded in the framework of effective field theory is to consider minute effects in the conventional field theory Lagrangians. The approach has been widely used and has several advantages: 1) it provides a very general setup without any attachment to a particular quantum gravity model and 2) it allows us to experimentally put bounds on the new physics. In particular, the search for Lorentz and CPT violation at low energies has been intensively studied from both the experimental and theoretical points of view.

Originally the framework of the standard model extension [1] was proposed with mass-dimension operators d lower than four and later generalized to include higher-order operators [2, 3]. Several higher-order effective models to describe departures from Lorentz symmetry have been proposed in the past years [4]. The effective field theory approach being a powerful tool, however, has to meet some constraints. In this work we emphasize the importance of having an S-matrix in the higher-order theory: (i) finite and (ii) unitary. For unitarity there have been some advances, see [5, 6, 7, 8].

Recently it has been shown that the Lorentz violation contained in quantum field theories leads to nontrivial modifications of the Källén–Lehmann representation and the LSZ reduction formalism [9]. This has led to modifications in the renormalization procedure, in the definition of the asymptotic Hilbert space and in general in the treatment for external-leg physics [10]; for other studies of the renormalization in Lorentz-breaking theories, see also [11]. A natural extension for these studies is to consider the nonminimal framework of Lorentz invariance violation, that is, when the Lorentz breaking is performed with higher-order operators [12].

2. The effective model

We focus on the higher-order Lorentz violating Lagrangian density

$$\mathcal{L} = \frac{1}{2}\partial_\mu \phi \partial^\mu \phi - \frac{1}{2}M^2\phi^2 + \bar{\psi}\left(i\partial\!\!\!/ - m\right)\psi + g_2\bar{\psi}n\!\!\!/(n\cdot\partial)^2\psi + g(\bar{\psi}\phi\psi)\,, \tag{1}$$

where the first two terms involve a scalar field and the last part is the Myers and Pospelov fermion model [4] with a Yukawa coupling interaction. The Planck mass suppressed constant g_2 parametrizes the higher-order Lorentz invariance violation, g is the Yukawa coupling and n^μ is a dimensionless four-vector defining a preferred reference frame.

The corresponding propagators in momentum space read

$$\Delta(p) = \frac{i}{p^2 - M^2}\,, \qquad S(p) = \frac{i}{p\!\!\!/ - m - g_2 n\!\!\!/(n\cdot p)^2}\,. \tag{2}$$

From now on we consider a purely time-like four-vector $n^\mu = (1,0,0,0)$. With this choice the scalar has the usual solutions but the number of fermion solutions increases being

$$\omega_1 = \frac{1-\sqrt{1-4g_2\mathcal{E}}}{2g_2}\,, \qquad \omega_2 = \frac{1-\sqrt{1+4g_2\mathcal{E}}}{2g_2}\,, \tag{3}$$

$$W_1 = \frac{1+\sqrt{1-4g_2\mathcal{E}}}{2g_2}\,, \qquad W_2 = \frac{1+\sqrt{1+4g_2\mathcal{E}}}{2g_2}\,, \tag{4}$$

with $\mathcal{E} = \sqrt{\vec{p}^2 + m^2}$. Note that the first two solutions are finite as $g_2 \to 0$ and the other two, the Lee-Wick modes, are singular [13, 14].

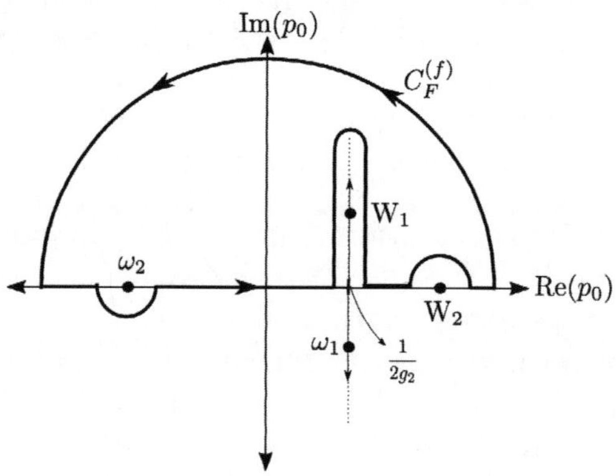

Figure 1. The Feynman contour $C_F^{(f)}$ for the fermion propagator.

As can be seen from the solutions above, at the energy $1/2g_2$ the solutions labelled "1" pick up an imaginary part and jump to the complex plane. To define the Feynman contour $C_F^{(f)}$ we use a heuristic argument. First, we consider the contour which can be continuously deformed to the standard curve at low energies when all the solutions are real and the Lorentz violation is removed. Second, we consider the contour that allows us to preserve the unitarity of the S-matrix [15]. In this way, the integration contour $C_F^{(f)}$ is defined to pass below the negative pole and above the three positive ones, as shown in Fig. 1.

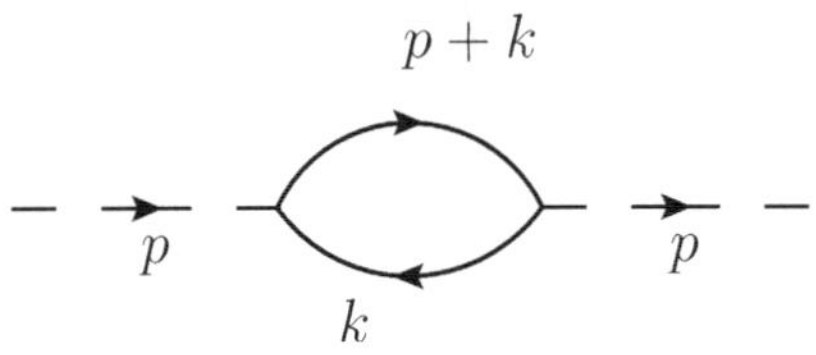

Figure 2. Self-energy loop graph in the scalar sector.

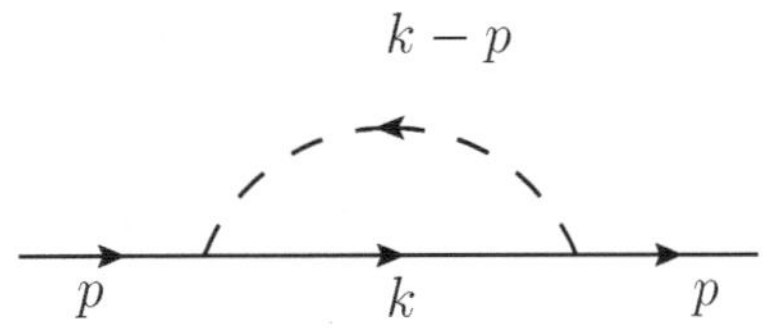

Figure 3. The fermion self-energy graph.

3. Two-point functions

As a first example of quantum corrections in our Yukawa-like model, we study the contribution with two external scalar legs depicted in Fig. 2 and represented by the integral

$$\Pi(p) = -\frac{g^2}{2}\phi(-p)\phi(p)\int\frac{d^4k}{(2\pi)^4}\frac{\mathrm{Tr}\left((Q_\mu\gamma^\mu+m)(R_\nu\gamma^\nu+m)\right)}{(Q^2-m^2)(R^2-m^2)}\,, \tag{5}$$

where we define

$$Q_\mu = k_\mu - g_2 n_\mu (n\cdot k)^2\,, \qquad R_\mu = k_\mu + p_\mu - g_2 n_\mu (n\cdot(k+p))^2\,. \tag{6}$$

We study the typical low-energy behavior of this contribution by expanding it into a Taylor series. After some algebra we find up to second order in p

$$\Pi(p) = -2g^2m^2q_0 - 2g^2p^2q_1 - 2g^2(n\cdot p)^2q_n\,, \tag{7}$$

where

$$q_0 = -\frac{i}{48\pi^2 g_2^2 m^2} + \frac{i}{48\pi^2}\left(6\gamma_E - 0.46 + 12i\pi - 18\ln\left(\frac{g_2 m}{2}\right)\right), \tag{8}$$

$$q_1 = -\frac{i}{2\pi^2}\left(i\pi - \ln\left(\frac{g_2 m}{2}\right) - \frac{1}{3}\right), \tag{9}$$

$$q_n = \frac{i}{\pi^2}\,. \tag{10}$$

Now we focus on the contribution of the fermion self-energy graph depicted in Fig. 3. We take $M = m$ and consider the fermion self-energy graph

$$\Sigma_2(p) = g^2\int\frac{d^4k}{(2\pi)^4}\frac{\not{Q}+m}{((k-p)^2-m^2)(Q^2-m^2)}\,. \tag{11}$$

After considering its Taylor expansion along the lines of the previous contribution, we obtain

$$\Sigma_2 = g^2\not{n}f_1^n + g^2mf_0 + g^2\not{p}f_1 + g^2m(n\cdot p)f_2^n + g^2\not{n}(n\cdot p)f_3^n + g^2p^2\not{n}f_6^n + g^2(n\cdot p)\not{p}f_4^n + g^2(n\cdot p)^2\not{n}f_5^n\,. \tag{12}$$

4. Mass renormalization

Let us start with the renormalized scalar Lagrangian

$$\mathcal{L} = \frac{1}{2}\partial_\mu\phi_R\partial^\mu\phi_R - \frac{1}{2}m_R^2\phi_R^2\,. \tag{13}$$

Consider the renormalized two-point function

$$(\Gamma^{(2)}_R)^{-1} = p^2 - m_R^2 + \Pi_R(p)\,, \tag{14}$$

where $\Pi_R(p) = \Pi(p) + \mathcal{O}(g^4)$. Let us write

$$\Pi_R(p) = p^2 A_\phi + m_R^2 B_\phi + (n\cdot p)^2 C_\phi\,, \tag{15}$$

where A_ϕ, B_ϕ and C_ϕ are the radiative corrections already computed in the expressions (8), (9), (10). We define a quantity $\bar{P}$ and $\bar{P}_\phi^2 = p^2 - M_{\text{ph}}^2 + \bar{y}(n\cdot p)^2$ with the appropriate structure to make the two-point function vanish as seen from Eq. (15). The above quantities M_{ph} and $\bar{y}$ are unknown constants we should find.

The new renormalization condition defined by $(\Gamma^{(2)}_R)^{-1}(\bar{P}_\phi^2) = 0$ at $\bar{P}_\phi^2 = 0$ gives the unknown constants

$$M_{\text{ph}}^2 = m_R^2 \frac{(1-B_\phi)}{(1+A_\phi)}\,, \qquad \bar{y} = \frac{C_\phi}{1+A_\phi}\,. \tag{16}$$

which are determined in terms of the radiative corrections.

Now we pass to the renormalization in the fermion sector and consider the renormalized Lagrangian

$$\mathcal{L} \;=\; i\bar{\psi}^R \partial\!\!\!/ \psi^R - m_R \bar{\psi}^R\psi^R + g_2 \bar{\psi}^R n\!\!\!/ (n\cdot\partial)^2 \psi^R + \delta_{g_2} \bar{\psi}^R p\!\!\!/ \psi^R\,, \tag{17}$$

with δ_{g_2} a counterterm. The renormalized two-point function is $(\Gamma^{(2)}_R)^{-1} = p\!\!\!/ - m_R + \Sigma_R$, with

$$\Sigma_R = \Sigma_2 + \delta_{g_2} n\!\!\!/ + \mathcal{O}(g^4)\,. \tag{18}$$

We can write the finite contribution to the renormalized two-point function up to second order in p with $\Sigma_R = p\!\!\!/ A_\psi + m_R B_\psi + n\!\!\!/ C_\psi$ where again the constants A_ψ, B_ψ and C_ψ have been determined in (12) through the expressions

$$\begin{aligned} A_\psi &= A^{(0)} + g_2 A^{(1)}(n\cdot p)\,, \qquad B_\psi = B^{(0)} + g_2 B^{(1)}(n\cdot p)\,, \\ C_\psi &= \frac{C^{(0)}}{g_2} + C^{(1)}(n\cdot p) + g_2 C^{(2)}(n\cdot p)^2 + g_2 C^{(3)} p^2\,. \end{aligned} \tag{19}$$

We work in the minimal subtraction scheme and hence we fix the divergent term to be $\delta_{g2} = \frac{ig^2}{4g_2\pi^2\epsilon}$. To find the pole we consider the ansatz $\bar{P}_\psi = p\!\!\!/ - \bar{m} + \bar{x} n\!\!\!/$ where $\bar{m} = m_{\text{ph}} + g_2 m_n (n\cdot p)$ and $\bar{x} = \frac{\bar{x}^{(0)}}{g_2} + \bar{x}^{(1)}(n\cdot p) + g_2\bar{x}^{(2)}(n\cdot p)^2 + g_2 \bar{x}^{(3)} p^2$. We include the large Lorentz violating terms in the pole extraction process which has been included explicitly in $\bar{x}$.

Considering the condition for $\bar{P}_\psi$ to be a pole, that is to say,

$$(\Gamma^{(2)}_R)^{-1}(\bar{P}_\psi = 0) = p\!\!\!/ - m_R + \Sigma_R = 0\,, \tag{20}$$

gives the six terms m_{ph}, m_n, $\bar{x}^{(0)}$, $\bar{x}^{(1)}$, $\bar{x}^{(2)}$, $\bar{x}^{(3)}$,

$$m_{\text{ph}} = m_R \frac{(1-B^{(0)})}{(1+A^{(0)})}\,, \qquad m_n = -m_R\left(\frac{(A^{(1)}(1-B^{(0)})}{(1+A^{(0)})^2} + \frac{B^{(1)}}{1+A^{(0)}}\right)\,. \tag{21}$$

and

$$\begin{aligned} \bar{x}^{(0)} &= \frac{C^{(0)}}{1+A^{(0)}}\,, \qquad \bar{x}^{(1)} = -\frac{A^{(1)}C^{(0)}}{(1+A^{(0)})^2} + \frac{C^{(1)}}{1+A^{(0)}}\,, \\ \bar{x}^{(2)} &= \frac{(A^{(1)})^2 C^{(0)}}{(1+A^{(0)})^3} - \frac{A^{(1)}C^{(1)}}{(1+A^{(0)})^2} + \frac{C^{(2)}}{1+A^{(0)}}\,, \qquad \bar{x}^{(3)} = \frac{C^{(3)}}{1+A^{(0)}}\,. \end{aligned} \tag{22}$$

This finishes the pole extraction in both sectors.

5. Summary

We considered the Myers-Pospelov-like higher-derivative extension of the Yukawa model which incorporates possible new physics from the Planck scale through dimension five operators coupled to a preferred four vector n^μ which breaks Lorentz symmetry. We have selected a particular configuration of Lorentz symmetry violation in which the preferred four-vector n^μ is purely timelike. This choice produces higher-order time-derivative terms and leads to extra solutions and new poles in the model. We have found and identified the poles associated to standard solutions and those ones corresponding to negative-metric states or Lee-Wick solutions. Some of these poles move on the real axis, as in the usual Yukawa model. However, above a specific energy called the critical energy some solutions become complex introducing an extra difficulty to define a consistent prescription for the contour of integration in the complex plane. To solve this problem we have analyzed the motion of the poles in the complex plane.

A central part of this work has been the study of mass renormalization in the presence of both Lorentz invariance violation and higher-order operators. We have computed the one-loop corrections in the model and shown how they lead to modifications in the renormalization procedure by pushing the pole mass to a sector involving other gamma matrices besides of $\not{p}$. This means basically that tree level physics is modified due to quantum corrections so that at asymptotic times one has to include corrections in the propagation of particles with the operators $\bar{P}_\phi$ and $\bar{P}_\psi$.

Acknowledgments

This work was partially supported by Conselho Nacional de Desenvolvimento Científico e Tecnológico (CNPq). The work by A. Yu. P. has been supported by the CNPq project No. 303783/2015-0. C. M. R. acknowledges support by the Chilean FONDECYT grant 1140781.

References

[1] Colladay D and Kostelecký V A 1997 *Phys. Rev.* D **55** 6760
[2] Kostelecký V A and Mewes M 2009 *Phys. Rev.* D **80** 015020
[3] Kostelecký V A and Mewes M 2013 *Phys. Rev.* D **88** 096006
[4] Myers R C and Pospelov M 2003 *Phys. Rev. Lett.* **90** 211601
[5] Reyes C M 2013 *Phys. Rev.* D **87** 125028
[6] Lopez-Sarrion J and Reyes C M 2013 *Eur. Phys. J.* C **73** 2391
[7] Klinkhamer F R and Schreck M 2011 *Nucl. Phys.* B **848** 90
[8] Schreck M 2014 *Phys. Rev.* D **90** 085025
[9] Potting R 2012 *Phys. Rev.* D **85** 045033
[10] Cambiaso M, Lehnert R and Potting R 2014 *Phys. Rev.* D **90** 065003
[11] Brito L C T, Fargnoli H G and Baêta Scarpelli A P 2013 *Phys. Rev.* D **87** 125023
[12] Nascimento J R, Petrov A Y and Reyes C M Renormalization in a Lorentz-violating model and higher-order operators *Preprint* arXiv:1706.01466 [hep-th]
[13] Lee T D and Wick G C 1969 *Nucl. Phys.* B **9** 209
[14] Lee T D and Wick G C 1970 *Phys. Rev.* D **2** 1033
[15] Reyes C M and Urrutia L F 2017 *Phys. Rev.* D **95** 015024

IOP Conf. Series: Journal of Physics: Conf. Series **952** (2018) 012018 doi:10.1088/1742-6596/952/1/012018

Vacuum Cherenkov radiation for Lorentz-violating fermions

M Schreck

Departamento de Física, Universidade Federal do Maranhão, 65080-805, São Luís, Maranhão, Brazil

E-mail: marco.schreck@ufma.br

Abstract. The current paper summarizes the content of a talk given on vacuum Cherenkov radiation emitted by Lorentz-violating fermions that are described in the context of the Standard-Model Extension. The decay rate will be obtained for various sets of controlling coefficients that will subsequently be constrained using cosmic-ray data.

1. Introduction

Ordinary Cherenkov radiation occurs whenever a particle carrying an electric charge moves through an optical medium with a speed that is larger than the phase velocity of light in that medium. Under this condition, the particle generates a large number of electric dipoles that start radiating their wave trains in phase such that they interfere constructively. The consequence is coherent radiation of high intensity that can be detected far away from the source. The latter is called Vavilov-Cherenkov radiation or most of the time simply Cherenkov radiation named after the Russian experimental physicists Vavilov and Cherenkov that discovered it in 1934. Only three years later, Frank and Tamm explained the effect theoretically in the context of classical electromagnetism. Cherenkov, Frank, and Tamm shared the Nobel Prize of 1958, which could be an explanation for why the name Vavilov is often omitted unjustifiedly. Cherenkov radiation is visible as a blue, glowing light in the water around nuclear reactors.

A Lorentz-violating background field turns the vacuum into an optical medium with nontrivial refractive index. The latter can depend on the energy of a propagating electromagnetic wave, which leads to dispersion. A direction dependence is possible, as well, and can result in birefringence, i.e., in a change of polarization of light as a function of propagation distance. Under certain circumstances, a charged particle loses energy by radiating photons when it moves through such a background field. Due to the fundamental characteristics shared with ordinary Cherenkov radiation, this process is referred to as vacuum Cherenkov radiation.

Vacuum Cherenkov radiation was probably mentioned in [1] for the first time in the context of modified static gravitational couplings. The process had been forgotten long since, until it had a revival at the end of the past millennium due to the rising interest in the field of Lorentz-symmetry violation. First, this increasing interest can be traced back to the emergence of a series of papers demonstrating how violations of Lorentz invariance can occur in quantum-gravity scenarios [2, 3]. Second, the construction of the minimal Standard-Model Extension (SME) as a model-independent framework parameterizing all deviations from Lorentz invariance in the Standard Model and General Relativity in a coordinate-invariant and gauge-invariant way [5, 6]

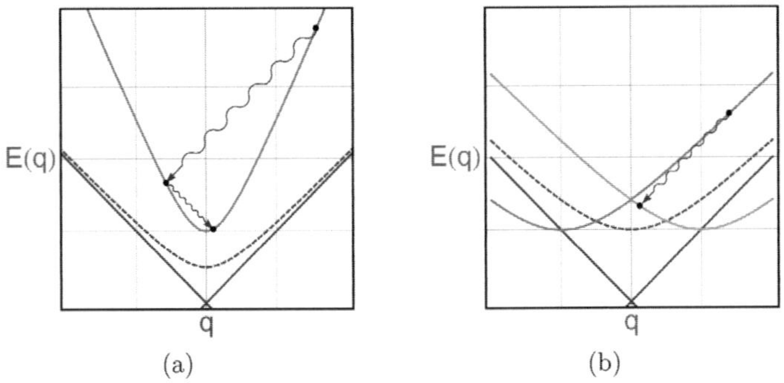

Figure 1: Standard null cone (violet), standard mass shell (dashed, blue) and modified mass shell(s) (red, orange) for the spin-degenerate case with $\mathring{c} = -1/2$ (a) and for the spin-nondegenerate case with $\mathring{b} = 1$ (b). Photon emission is indicated by green, wiggly lines.

finally allowed for a systematic description of this effect. Since then, plenty of papers studying vacuum Cherenkov radiation have appeared with a small excerpt of this list given by [7–10]. Most of these works were carried out in the framework of either a modified electrodynamics or for modified photons where only recently papers have been published on gravitational Cherenkov radiation [11] and vacuum Cherenkov radiation for pions [12] and for a modified electroweak sector [13]. Until lately, vacuum Cherenkov radiation for Lorentz-violating fermions has evaded investigation. This was the motivation for the article [4] that is on the focus.

2. Fermion sector of the minimal Standard-Model Extension

Within the minimal SME, Lorentz-violating fermions are described by including all power-counting renormalizable Lorentz-violating terms into the Dirac Lagrange density, cf. Eq. (1) in [14]. Each of these contributions is decomposed into controlling coefficients and a field operator appropriately contracted such that the term is an observer Lorentz scalar. The controlling coefficients can be interpreted as background fields permeating the vacuum and resulting in velocity- and direction-dependent physics. The b, d, g, and H coefficients are spin-nondegenerate, i.e., they produce two distinct dispersion relations for particles and antiparticles, respectively. The dispersion relation depends on the spin projection with respect to the quantization axis.

3. Modified kinematics

Basically, there are two different types of processes. The first involves an incoming fermion of a certain spin projection that emits a photon such that the spin projection is conserved. It is reasonable to study its kinematics based on plots of modified mass shells and null cones. As an example, we analyze the isotropic coefficient $\mathring{c}$, cf. Fig. 1a. The modified mass shell evidently has a larger curvature in comparison to the standard mass shell. Considering a point on the mass shell far enough away from the origin, this property allows the modified fermion to emit a photon, lose energy, and reverse the direction of its momentum. An emitted photon corresponds to a line parallel to the standard null cone. The emission of several photons is possible as long as the fermion energy is large enough whereupon its energy decreases steadily and the direction of its momentum reverses for each emission. Using energy conservation, it is shown that the process is forbidden for the dimensionful coefficients, i.e., for a, b, and H.

The second type of process can only occur once the fermion is modified by spin-nondegenerate coefficients. Here, the spin projection of the initial fermion changes after photon emission, i.e., a spin flip occurs. In the literature, such a process is often called "helicity decay." As an example, we consider an isotropic framework with a nonzero $\mathring{b}$, cf. Fig. 1b. There are two different branches of the mass shell that intersect themselves and this intersection is visible as a cusp. With the fermion being on a certain branch, a photon can be emitted such that the final state lies on the other branch. This change of the branch corresponds to a spin flip. Helicity decays have certain characteristics that are absent for the previous type of process. First, multiple photons cannot be emitted, as this would require a further impossible change of the branch. Second, the direction of the fermion momentum is not reversed. After all, the required change of helicity for photon emission is induced by the spin flip. Energy conservation shows that helicity decays are allowed for all fermion coefficients except a. Hence, the latter coefficients are the only ones that do not render vacuum Cherenkov radiation possible.

4. Modified dynamics

The great success of the SME rests on the fact that it is based on field theory, i.e., it does not only allow for describing a modified kinematics, but enables studies of a modified dynamics, as well. This possibility is granted by the modified Lagrange density and Dirac equation. The latter can be solved for the dispersion relations and the spinor solutions describing asymptotic fermions in field theory. These solutions were derived in [16] for various sets of both minimal and nonminimal coefficients, which is why that paper can be considered as the theoretical base for the work [4]. The spinor solutions are needed to calculate the matrix element squared at tree level with an initial fermion of spin s and momentum q to a final fermion of spin s' and momentum $p = q - k$:

$$|\mathcal{M}^{(s,s')}|^2 = 4\pi\alpha\,\mathrm{Tr}[\Lambda^{(s)}(q)\Gamma^{\mu}\Lambda^{(s')}(q-k)\Gamma^{\nu}]\Pi_{\mu\nu}(k)\,, \tag{1a}$$

$$\Pi_{\mu\nu}(k) \equiv \sum\nolimits_{\lambda=1,2} \varepsilon_{\mu}^{(\lambda)}(k)\varepsilon_{\nu}^{(\lambda)}(k)\,, \tag{1b}$$

$$\Gamma^{\nu} = \gamma^{\nu} + c^{(4)\mu\nu}\gamma_{\mu} + d^{(4)\mu\nu}\gamma^5\gamma_{\mu} + e^{(4)\nu}1_4 + \mathrm{i}f^{(4)\nu}\gamma^5 + (1/2)g^{\varrho\sigma\nu}\sigma_{\varrho\sigma}\,, \tag{1c}$$

$$\Lambda_{ab}^{(s)}(q) \equiv u_a^{(s)}(q)\bar{u}_b^{(s)}(q)\,, \tag{1d}$$

with the fine-structure constant $\alpha = e^2/(4\pi)$, the unit matrix 1_4 in spinor space, the chiral Dirac matrix $\gamma^5 = \mathrm{i}\gamma^0\gamma^1\gamma^2\gamma^3$, and $\sigma^{\mu\nu} = (\mathrm{i}/2)[\gamma^{\mu}, \gamma^{\nu}]$ where γ^{μ} are the standard Dirac matrices. Furthermore, $\Pi_{\mu\nu}$ is the sum over the physical photon polarization tensors with the polarization vectors $\varepsilon_{\mu}^{(\lambda)}$ for a transverse polarization mode λ. The trace in spinor space is carried out over a train of quantities. The minimal coupling of the modified Dirac theory to photons produces a modified vertex Γ^{ν} that involves all controlling coefficients contracted with a derivative in the Lagrange density. The quantities $\Lambda^{(s)}$ are matrices formed from a particle spinor $u^{(s)}$ and its Dirac-conjugated counterpart $\overline{u}^{(s)} \equiv u^{(s)\dagger}\gamma^0$. It was also demonstrated in [16] that these matrices can be neatly computed directly from the modified propagators of the theory. The matrix element squared shown above is the field theoretical object needed to derive the decay rate for a helicity decay. However, once spin-degenerate coefficients are involved, the significant object is the following:

$$|\mathcal{M}|^2 = \frac{1}{2}\sum\nolimits_{s,s'=\pm} |\mathcal{M}^{(s,s')}|^2\,. \tag{2}$$

As we consider the initial particle beam to be unpolarized, the initial spin s is averaged over. Since the final-particle state is unclear, the two spin projections s' must be summed over. Finally, we also take into account Lorentz-violating contributions that contain additional time derivatives. These derivatives are known to spoil the proper time evolution of asymptotic fermion states, which is why they must be removed by a suitable transformation in spinor space [15].

The matrix element squared delivers the probability for a certain process to occur with explicitly given final-particle momenta. However, quantum physics makes the final-particle state undetermined, which is why all possible momentum configurations have to be integrated over. This integration is six-dimensional and can immediately be reduced to three dimensions on the base of momentum conservation. The remaining three integrations must be performed by taking into account energy conservation. For an isotropic framework, we can parameterize the

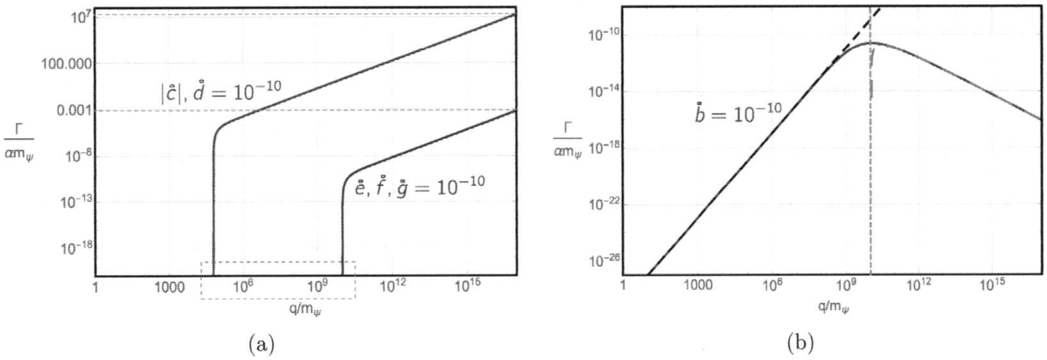

Figure 2: Double-logarithmic plot of decay rates $\Gamma/(\alpha m_\psi)$ for vacuum Cherenkov radiation as functions of q/m_ψ without a spin flip (a) and with a spin flip of the initial fermion (b).

photon phase space with spherical polar coordinates and the azimuthal angle can be integrated over directly. The integration over the polar angle sets this angle to a particular value that is in accordance with energy conservation. The final remaining integration must be carried out over the magnitude of the photon momentum that runs from a vanishing momentum to a finite maximum k_{max}. In the Lorentz-invariant case, the latter k_{max} vanishes, as there is no phase space available for this process to occur.

5. Decay rates and threshold momenta

The explicit result for the matrix element squared is quite complicated even for simple coefficient choices. Hence, the phase space is integrated over numerically with computer algebra. For the spin-degenerate, isotropic coefficients, the results are shown in Fig. 2a. The dimensionless coefficients are chosen to be 10^{-10} for the plot. For such tiny coefficients, the curves for $\mathring{c}$, $\mathring{d}$ and $\mathring{e}$, $\mathring{f}$, $\mathring{g}$ cannot be distinguished from each other. An important property of the spin-conserving process is that it has a threshold, i.e., the decay rate goes to zero when the particle momentum approaches a certain value from above. Furthermore, for large momenta, the decay rate increases linearly with momentum. The decay rates for the second set of coefficients are suppressed in comparison to the decay rates for $\mathring{c}$, $\mathring{d}$ by an additional power in Lorentz violation.

It is also evident that the threshold energies for both groups of coefficients differ from each other. At leading order in Lorentz violation, the threshold momenta are given by

$$q_{\mathring{c}}^{\text{th}} = \frac{1}{2}\sqrt{\frac{3}{2}}\frac{m_\psi}{\sqrt{-\mathring{c}}}, \quad q_{\mathring{d}}^{\text{th}} = \frac{1}{2}\sqrt{\frac{3}{2}}\frac{m_\psi}{\sqrt{\pm\mathring{d}}}, \quad q_{\mathring{e}}^{\text{th}} = \frac{m_\psi}{\pm\mathring{e}}, \quad q_{\mathring{g}}^{\text{th}} = \frac{m_\psi}{\pm\mathring{g}}. \tag{3}$$

The threshold momenta are proportional to the fermion mass due to dimensional reasons. They depend on the inverse of the coefficients, which indicates that the process is forbidden for vanishing Lorentz violation. Furthermore, the coefficient $\mathring{c}$ must be chosen negative to render the decay possible. For the spin-nondegenerate $\mathring{d}$, $\mathring{g}$ both signs of coefficients are allowed, as there are two dispersion relations whose first-order terms in Lorentz violation simply differ by a sign. Last but not least, both signs are also permissible for $\mathring{e}$ although the latter is spin-degenerate. The lower signs for $\mathring{d}$, $\mathring{e}$, and $\mathring{g}$ are always understood to hold for negative SME coefficients.

The decay rates for the helicity decays of the dimensionful isotropic coefficients have completely different properties and for comparison, the result for a nonzero $\mathring{b}$ is presented in Fig. 2b. Here, the decay does not have a threshold, but it is highly suppressed by Lorentz violation. Furthermore, the rate has a maximum beyond which it decreases again, as for large momenta it becomes more and more difficult for the fermion spin to flip.

6. Phenomenology

We consider an ultra-high energetic particle propagating to the Earth after having been accelerated, e.g., in the shock front of a supernova remnant. At least for the c, d coefficients, the radiated-energy rates are large enough such that the particle loses its energy fraction above the threshold after few centimeters, i.e., the decay is very efficient for these coefficients. Hence, when a cosmic ray is detected on Earth, its energy can be assumed to be smaller than the threshold.

To obtain constraints, we need a cosmic-ray event with known energy observed on Earth. A suitable event of $E = 212\,\mathrm{EeV}$ was detected by the Pierre-Auger observatory. A conservative estimate is obtained by assuming that the primary was an iron nucleus with nucleon number $N = 56$, i.e., each nucleon had an energy of E/N. We consider the quark structure of the nucleons where vacuum Cherenkov radiation can be emitted from the real up and down quarks only. A sophisticated analysis would require to take into account the parton distribution functions. However, we follow the simple, but reasonable estimate that each of these quarks carries a fraction $r = 0.1$ of the total energy of the nucleon. We then obtain the following set of bounds on the up-quark coefficients from the threshold energies:

$$-3\times 10^{-23} < \mathring{c}^{\mathrm{u}} - (3/8)(\mathring{f}^{\mathrm{u}})^2\,, \tag{4a}$$

$$-3\times 10^{-23} < \mathring{d}^{\mathrm{u}} < 3\times 10^{-23}\,, \tag{4b}$$

$$-9\times 10^{-12} < \mathring{e}^{\mathrm{u}}, \mathring{g}^{\mathrm{u}} < 9\times 10^{-12}\,. \tag{4c}$$

Note that the f coefficients squared can be moved to the c coefficients by a field redefinition. The constraint on the c coefficients is one-sided only, as vacuum Cherenkov radiation can only occur for a negative coefficient.

7. Acknowledgments

The author thanks the organizers of the workshop for the invitation and their hospitality. Furthermore, he acknowledges financial support via the grant FAPEMA/POS-GRAD/03978/15.

References

[1] Beall E F 1970 *Phys. Rev.* D **1** 961
[2] Kostelecký V A and Samuel S 1989 *Phys. Rev.* D **39** 683
[3] Gambini R and Pullin J 1999 *Phys. Rev.* D **59** 124021
[4] Schreck M 2017 Vacuum Cherenkov radiation for Lorentz-violating fermions *Preprint* arXiv:1702.03171 [hep-ph]
[5] Colladay D and Kostelecký V A 1998 *Phys. Rev.* D **58** 116002
[6] Kostelecký V A 2004 *Phys. Rev.* D **69** 105009
[7] Lehnert R and Potting R 2004 *Phys. Rev.* D **70** 125010
[8] Kaufhold C and Klinkhamer F R 2007 *Phys. Rev.* D **76** 025024
[9] Altschul B 2007 *Phys. Rev. Lett.* **98** 041603
[10] Klinkhamer F R and Schreck M 2008 *Phys. Rev.* D **78** 085026
[11] Kostelecký V A and Tasson J D 2015 *Phys. Lett.* B **749** 551
[12] Altschul B 2016 *Phys. Rev.* D **93** 105007
[13] Colladay D, Noordmans J P and Potting R 2017 *Phys. Rev.* D **96** 035034
[14] Kostelecký V A and Mewes M 2013 *Phys. Rev.* D **88** 096006
[15] Colladay D and Kostelecký V A 2001 *Phys. Lett.* B **511** 209
[16] Reis J A A S and Schreck M 2017 *Phys. Rev.* D **95** 075016

IOP Conf. Series: Journal of Physics: Conf. Series **952** (2018) 012019 doi:10.1088/1742-6596/952/1/012019

Repercussions of dimension five nonminimal couplings in the electroweak model

Victor E Mouchrek-Santos and Manoel M Ferreira Jr

Departamento de Física, Universidade Federal do Maranhão, Campus Universitário do Bacanga, São Luís - MA, 65080-805 - Brazil

E-mail: victor_mouchrek@hotmail.com, manojr.ufma@gmail.com

Abstract. We propose two possibilities of CPT-odd, dimension five and Lorentz-violating (LV) nonminimal couplings in the Electroweak model. These gauge-invariant terms couple a fixed LV 4-vector to the electroweak field strengths. After finding the LV contributions to the electroweak currents, we evaluate the decay rate for the vector mediators W and Z. Using the known experimental uncertainty in these decay rates, upper bounds at the level of 10^{-15} $(\mathrm{eV})^{-1}$ and 10^{-14} $(\mathrm{eV})^{-1}$ are imposed on the magnitude of the proposed nonminimal interactions.

1. Introduction

The most general effective theory considering the explicit violation of Lorentz and CPT symmetry is the minimal standard model extension (mSME) [1], which is an extension of the $SU(3)\times SU(2)\times U(1)$ standard model, exhibiting terms violating Lorentz and CPT symmetries in all of its sectors: lepton, quark, Yukawa, Higgs and gauge. Nonminimal LV interactions have also been examined in an extension of the SME encompassing higher-order derivatives in both the gauge [2] and the fermion sectors [3]. Some models possessing higher-dimension operators [4] and higher derivatives [5] have also been proposed and developed. Nonminimal interactions have been a topical issue in the latest years, mainly in the fermion and electromagnetic sectors [7]. A systematic investigation on nonminimal couplings (NMCs) of dimension five and six was recently proposed in Ref. [6].

In the Glashow-Salam-Weinberg electroweak model (GSW), with a $SU(2)_L \times U(1)_Y$ gauge structure spontaneously broken via the Higgs mechanism, the vector bosons, $W^{\pm}$, Z^0 and γ work as mediators of the interactions, being introduced via minimal coupling to the matter fields. In this theory, left-handed leptons (L_l) are represented by isodublets and right-handed leptons (R_l) are isosinglets,

$$L_l = \begin{bmatrix} \psi_{\nu_l} \\ \psi_l \end{bmatrix}_L = \frac{1-\gamma_5}{2}\begin{bmatrix} \psi_{\nu_l} \\ \psi_l \end{bmatrix}, \; R_l = (\psi_l)_R = \left(\frac{1+\gamma_5}{2}\right)\psi_l, \tag{1}$$

and $l = 1,2,3$ is the lepton flavor label: $\psi_l = (e,\mu,\tau)$. The part of the electroweak Lagrangian, in which the leptons interact directly with the gauge fields, is $\mathcal{L}_{EW} = \mathcal{L}_{gauge} + \mathcal{L}_{lepton}$, where

$$\mathcal{L}_{gauge} = -\frac{1}{4}\mathbf{W}_{\mu\nu}\cdot\mathbf{W}^{\mu\nu} - \frac{1}{4}B_{\mu\nu}B^{\mu\nu}, \quad \mathcal{L}_{lepton} = \bar{L}_l\gamma^{\mu}iD_{\mu}L_l + \bar{R}_l\gamma^{\mu}iD_{\mu}R_l, \tag{2}$$

with $\mathbf{W}_\mu = (W_\mu^1, W_\mu^2, W_\mu^3)$ being a four-vector gauge field which is a three-vector in isospin space, and B_μ a gauge four-vector field. The associated field strengths are

$$\mathbf{W}_{\mu\nu} = \partial_\mu \mathbf{W}_\nu - \partial_\nu \mathbf{W}_\mu + g\mathbf{W}_\mu \times \mathbf{W}_\nu, \quad B_{\mu\nu} = \partial_\mu B_\nu - \partial_\nu B_\mu. \tag{3}$$

The covariant derivative involves both gauge fields,

$$D_\mu = \partial_\mu - ig\mathbf{T}\cdot\mathbf{W}_\mu - i\frac{g'}{2}YB_\mu. \tag{4}$$

where g, g' are the coupling constants, $\mathbf{T} = (T_1, T_2, T_3)$ stands for the generators of the group $SU(2)_L$, and Y is the generator of $U(1)_Y$ group. The generators and isovector can be also written as $\mathbf{T} = (T_+, T_3, T_-)$, $\mathbf{W}_\alpha = (W_\alpha^{(+)}/\sqrt{2}, W_\mu^3, W_\alpha^{(-)}/\sqrt{2})$, where $T_\pm = \sigma_x/2 \pm i(\sigma_y/2)$, $T_3 = \sigma_z/2$, $W_\mu^{(\pm)} = \frac{1}{\sqrt{2}}\left(W_\mu^1 \mp iW_\mu^2\right)$ and $W_\mu^3 = \sin\theta A_\mu + \cos\theta Z_\mu$ with $\sigma_x, \sigma_y, \sigma_z$ being the Pauli matrices, $A_\mu$and Z_μ denoting the photon field and the neutral intermediate boson respectively, and θ standing for the weak mixing angle.

The electroweak lepton sector of the mSME [1] is composed of CPT-even and CPT-odd terms,

$$\mathcal{L}_{lep}^{even} = (c_L)_{\mu\nu AB}\bar{L}_A\gamma^\mu iD^\nu L_B + (c_R)_{\mu\nu AB}\bar{R}_A\gamma^\mu iD^\nu R_B, \tag{5}$$
$$\mathcal{L}_{lep}^{odd} = -(a_L)_{\mu AB}\bar{L}_A\gamma^\mu L_B - (a_R)_{\mu AB}\bar{R}_A\gamma^\mu R_B, \tag{6}$$

where $A, B = 1, 2, 3$ are the lepton flavor labels and the constant coefficients $(c_I)_{\mu\nu AB}$ and $(a_I)_{\mu AB}$, with $I = L, R$, produce the breakdown of Lorentz symmetry. LV studies in the electroweak sector were initially developed in connection with meson decays $(\pi^- \to \mu^- + \bar{\nu}_\mu)$, where the LV effects were considered at the level of the Feynman propagator of the W boson [8], $\langle W^{\mu\dagger}W^\nu\rangle = -i\left(g^{\mu\nu} + \chi^{\mu\nu}\right)/M_W^2$, implying upper bounds of 1 part in 10^4. Detailed contributions to the W propagator were more explicitly considered in Ref. [9], with several implications on the nuclear β decays [10], [11]. It was also examined the possibility of LV electroweak terms making feasible forbidden processes $(Z_0 \longrightarrow \gamma+\gamma)$ [12] or modifying reactions such as $\gamma + e \to W + \nu_e$, $\gamma + \gamma \to W + W$ [13]. Lepton flavor violating decays triggered by renormalizable and nonrenormalizable (dimension five) terms belonging to the Higgs sector were recently considered as well [14].

In the present work, we discuss a different way of considering Lorentz violation in the electroweak model: a NMC in the lepton sector [16]. In this way, we introduce two possibilities of CPT-odd LV nonminimal interactions in the Electroweak sector, the first one being proposed in the $U(1)_Y$ sector of the GSW (Glashow-Salam-Weinberg) model, while the second is considered in its $SU(2)_L$ sector, both as extensions of the covariant derivative. After determining the terms engendered in the interaction Lagrangian, we evaluate the Lorentz-violating corrections to the decay rates of the following mediators: $Z_0 \to \bar{l} + l$ and $W^- \to l + \bar{\nu}_l$. With these results and the experimental uncertainty in the measurements, we impose upper limits on the magnitude of the LV parameters, achieving upper bounds as tight as 10^{-6} (GeV)$^{-1}$.

2. A nonminimal coupling in the $\mathrm{U}(1)_Y$ sector of the GSW model

Gauge invariant NM interactions in the electroweak sector can be proposed in the context of the covariant derivative (4). A first possibility, in the $U(1)_Y$ sector of the GSW model, is to include a dimension five term in it,

$$D_\mu = \partial_\mu - ig\mathbf{T}\cdot\mathbf{W}_\mu - i\frac{g'}{2}YB_\mu + ig_2'YB_{\mu\nu}C^\nu, \tag{7}$$

involving C^ν as a fixed 4-vector that establishes a preferred direction in spacetime and violates Lorentz symmetry. Replacing such a derivative in Lagrangian (2), the nonminimal coupling yields additional electromagnetic and neutral LV interactions, $\mathcal{L}_{LV(1)} = J^{(l)\nu}_{EM(LV)} A_\nu + J^{(l)\nu}_{0(LV)} Z_\nu$, given explicitly as

$$J^{(l)\nu}_{EM(LV)} = \frac{g_2'}{2}\cos\theta \left[\bar{\psi}_{\nu_l}\gamma^\mu (1-\gamma_5)\psi_{\nu_l}\right] C^\nu \partial_\mu - \frac{g_2'}{2}\cos\theta \left[\bar{\psi}_{\nu_l}\gamma^\nu (1-\gamma_5)\psi_{\nu_l}\right] C^\mu \partial_\mu + g_2' \cos\theta [j_1^\mu C^\nu \partial_\mu] - g_2' \cos\theta \left[j_1^\nu C^\mu \partial_\mu\right], \tag{8}$$

$$J^{(l)\nu}_{0(LV)} = -\frac{g_2'}{2}\sin\theta \left[\bar{\psi}_{\nu_l}\gamma^\mu (1-\gamma_5)\psi_{\nu_l}\right] C^\nu \partial_\mu + \frac{g_2'}{2}\sin\theta \left[\bar{\psi}_{\nu_l}\gamma^\nu (1-\gamma_5)\psi_{\nu_l}\right] C^\mu \partial_\mu - g_2' \sin\theta \left[j_1^\mu C^\nu \partial_\mu\right] + g_2' \sin\theta \left[j_1^\nu C^\mu \partial_\mu\right], \tag{9}$$

with $j_1^\mu(x) = \bar{\psi}_l(x)\gamma^\mu(3+\gamma_5)\psi_l(x)/2$. These expressions are useful to show the processes that are directly affected, at tree-level, by the nonminimal derivative (7). We now examine the effect of this nonminimal coupling on the Z_0 decay in a pair lepton and antilepton, $Z_0 \to \bar{l} + l$, evaluating the contributions implied to the decay rate. The total neutral current that contributes for this process is

$$\left(J_0^{(l)\mu} + J^{(l)\mu}_{0(LV)}\right) Z_\mu = -\frac{g}{4\cos\theta}\bar{\psi}_l(x)\gamma^\mu\left(g_V' - \gamma_5\right)\psi_l(x) Z_\mu(x) - g_2'\sin\theta\left[j_1^\eta C^\mu \partial_\eta Z_\mu(x)\right] + g_2'\sin\theta\left[j_1^\mu C^\lambda \partial_\lambda Z_\mu(x)\right], \tag{10}$$

where $g_V' = 1 - 4\sin^2\theta$. The first term in the above equation is the usual Lorentz invariant contribution, while the second and third terms stemming from Eq. (9) encoded the Lorentz-violating effects.

The scattering matrix for such a process is

$$S = -i\int d^4x \left(J_0^{(l)\mu} + J^{(l)\mu}_{0(LV)}\right) Z_\mu = S_0 + S_{LV(1)} + S_{LV(2)}, \tag{11}$$

where the zero order and first order contributions in the LV parameters are

$$S_0 = i\frac{g}{4\cos\theta}\int d^4x\, \bar{\psi}_l(x)\gamma^\mu\left(g_V' - \gamma_5\right)\psi_l(x) Z_\mu(x), \tag{12}$$

$$S_{LV(1)} = i g_2'\sin\theta \int d^4x \left[j_1^\eta(x) C^\mu \partial_\eta Z_\mu(x)\right], \tag{13}$$

$$S_{LV(2)} = -i g_2'\sin\theta \int d^4x \left[j_1^\mu(x) C^\lambda \partial_\lambda Z_\mu(x)\right]. \tag{14}$$

In order to evaluate these elements, we propose plane wave expansions, $Z^0_\mu(x) = N_k \varepsilon_\mu(k,\lambda)\exp(-ik\cdot x)$, $\psi_l(x) = N_q u_l(q,s)\exp(-iq\cdot x)$, $\psi_{\bar{l}}(x) = N_{q'} v(q',s')\exp(iq'\cdot x)$, where k, q, q' stand for the 4-momentum of the Z^0 boson and the emerging leptons, respectively, and $N_q = (2Vq_0)^{-1/2}$.

After a long evaluation, the total decay rate for the decay, $Z_0 \to \bar{l} + l$, is found to be:

$$\Gamma_{ll} = \frac{g^2(8M_Z)}{1536\pi\cos^2\theta}\left\{\left[\left(g_V'^2 + 1\right) - 6g_V'^2\frac{m_l^2}{M_Z^2}\right] - \frac{g_2'\sin 2\theta}{g}(C\cdot k)\left[\left(3g_V' - 2\right) - 27g_V'\frac{m_l^2}{M_Z^2}\right]\right\} \times \Theta(M_Z - 2m_l). \tag{15}$$

We now use $k^2 = M_Z^2$ and $C\cdot k = C_0 M_Z$. As the Z_0 mass ($M_Z = 9.1\times 10^{10}$ eV) is much larger than lepton masses, we can neglect the mass ratios for the electron, muon and tau

$(m_e^2/M_Z^2 \simeq 2\times 10^{-11},\ m_\mu^2/M_Z^2 \simeq 10^{-6}, m_\tau^2/M_Z^2 \simeq 4\times 10^{-4})$, which are smaller than the experimental uncertainty in decay rate measurements. Thus, the result is written as

$$\Gamma_{ll} = \frac{g^2\left(g_V'^2+1\right)M_Z}{192\pi\cos^2\theta}\left[1-8\times\left|g_2'C_0\right|M_Z\right]\Theta\left(M_Z-2m_l\right), \tag{16}$$

with the LV contribution appearing as a direct correction to the usual decay rate. We have used $g = e/\sin\theta$, $g_V' = 1-4\sin^2\theta$, $\sin^2\theta = 0.23$. In accordance with Ref. [15], the Z_0 decay rate (considering lepton universality) is $\Gamma_{ll} = (83.985\pm 0.086)$ MeV, or $\Gamma_{ll} = 83.985(1\pm 0.001)$ MeV, so that the experimental uncertainty is of 1 part in 10^3. We thus impose $8\left|g_2'C_0\right|M_Z < 1.0\times 10^{-3}$, which leads to the upper bound $\left|g_2'C_0\right| < 1.3\times 10^{-15}\ (\text{eV})^{-1}$, that is,

$$\left|g_2'C_0\right| < 1.3\times 10^{-6}\ (\text{GeV})^{-1}. \tag{17}$$

3. A nonminimal coupling in the SU$(2)_L$ sector of the GSW model

A gauge invariant dimension five nonminimal coupling in the $SU\,(2)_L$ sector of the GSW model can be proposed as

$$D_\mu = \partial_\mu - ig\mathbf{T}\cdot\mathbf{W}_\mu - i\frac{g'}{2}YB_\mu + ig_3'\left(\mathbf{T}\cdot\mathbf{W}_{\mu\nu}\right)V^\nu. \tag{18}$$

where V^ν is a fixed 4-vector that establishes a preferred direction in spacetime and violates Lorentz symmetry. The coupling term, $\bar{L}_l\gamma^\mu i\left(ig_3'\mathbf{T}\cdot\mathbf{W}_{\mu\nu}V^\nu\right)L_l$ embraces the following interactions at tree-level, $\mathcal{L}_{LV(2)} = \mathcal{J}_{+(LV)}^{(l)\nu}W_\nu^{(+)} + \mathcal{J}_{-(LV)}^{(l)\nu}W_\nu^{(-)} + \mathcal{J}_{0(LV)}^{(l)\nu}Z_\nu$, involving the vector bosons, in which the related currents are

$$\mathcal{J}_{+(LV)}^{(l)\nu} = -\frac{g_3'}{2\sqrt{2}}\bar{\psi}_{\nu_l}\gamma^\mu\left(1-\gamma_5\right)\psi_l V^\nu\partial_\mu + \frac{g_3'}{2\sqrt{2}}\bar{\psi}_{\nu_l}\gamma^\nu\left(1-\gamma_5\right)\psi_l V^\mu\partial_\mu, \tag{19}$$

$$\mathcal{J}_{-(LV)}^{(l)\nu} = -\frac{g_3'}{2\sqrt{2}}\bar{\psi}_l\gamma^\mu\left(1-\gamma_5\right)\psi_{\nu_l} V^\nu\partial_\mu + \frac{g_3'}{2\sqrt{2}}\bar{\psi}_l\gamma^\nu\left(1-\gamma_5\right)\psi_{\nu_l} V^\mu\partial_\mu, \tag{20}$$

$$\mathcal{J}_{0(LV)}^{(l)\nu} = -\frac{g_3'\cos\theta}{4}\left\{\bar{\psi}_{\nu_l}\gamma^\mu\left(1-\gamma_5\right)\psi_{\nu_l}V^\nu\partial_\mu - \bar{\psi}_l\gamma^\mu\left(1-\gamma_5\right)\psi_l V^\nu\partial_\mu - \bar{\psi}_{\nu_l}\gamma^\nu\left(1-\gamma_5\right)\psi_{\nu_l}V^\mu\partial_\mu + \bar{\psi}_l\gamma^\nu\left(1-\gamma_5\right)\psi_l V^\mu\partial_\mu\right\}. \tag{21}$$

The current, $J_{-(LV)}^{(l)\mu}$, given by Eq. (20), affects the processes mediated by the W^- particle, including the decay $W^-\to l+\bar{\nu}_l$. The total electroweak current that contributes to this process is

$$\left(J_-^{(l)\mu} + \mathcal{J}_{-(LV)}^{(l)\mu}\right)W_\mu^{(-)} = \frac{1}{2\sqrt{2}}\left\{gj_2^\mu W_\mu^{(-)}(x) - g_3'[j_2^\eta V^\mu\partial_\eta W_\mu^{(-)}(x)] + g_3'[j_2^\mu V^\lambda\partial_\lambda W_\mu^{(-)}(x)]\right\}, \tag{22}$$

where $j_2^\mu(x) = \bar{\psi}_l(x)\gamma^\mu\left(1-\gamma_5\right)\psi_{\nu_l}$, and the first term is the usual Lorentz invariant contribution. The scattering matrix for the process $\left(W^-\to l+\bar{\nu}_l\right)$, at leading order, can be written as $\mathcal{S} = -i\int d^4x\left(J_-^{(l)\mu} + \mathcal{J}_{-(LV)}^{(l)\mu}\right)W_\mu^{(-)}$, that implies $\mathcal{S} = \mathcal{S}_0 + \mathcal{S}_{LV(1)} + \mathcal{S}_{LV(2)}$, with

$$\mathcal{S}_0 = -i\frac{g}{2\sqrt{2}}\int d^4x\ \left[j_2^\mu(x)W_\mu^{(-)}(x)\right], \tag{23}$$

$$\mathcal{S}_{LV(1)} = i\frac{g_3'}{2\sqrt{2}}\int d^4x\ \left[j_2^\eta(x)V^\mu\partial_\eta W_\mu^{(-)}(x)\right], \tag{24}$$

$$\mathcal{S}_{LV(2)} = -i\frac{g_3'}{2\sqrt{2}}\int d^4x\ \left[j_2^\mu(x)V^\lambda\partial_\lambda W_\mu^{(-)}(x)\right]. \tag{25}$$

The total decay rate for the decay, $W^- \to l + \bar{\nu}_l$, is read as

$$\Gamma = \frac{g^2}{48\pi} M_W \left[1 + (g_3' V_0) \frac{5M_W}{4g} \right] \Theta(M_W - m_l), \tag{26}$$

where $V \cdot k = V_0 M_W$ for the rest frame of the W^- mediator, and we have neglected the contributions in $m_l^2/M_W^2, m_l^4/M_W^4$. Considering that the experimental uncertainty in the measures of this decay is at the level of $\sim 4.0 \times 10^{-2}$, and using $g = e/\sin\theta$, $\sin^2\theta = 0.23$, we impose $7\,(g_3' V_0)\, M_W < 4.0 \times 10^{-2}$, yielding $|g_3' V_0| < 7 \times 10^{-14}\ (\text{eV})^{-1}$, or

$$|g_3' V_0| < 7 \times 10^{-5}\ (\text{GeV})^{-1}. \tag{27}$$

As the current (19), involving the mediator W^+, is analogue to the current (20), we conclude that these latter developments equally hold to the decay $W^+ \to \bar{l} + \nu_l$, which becomes constrained by a bound similar to Eq. (27).

4. Final remarks and Conclusion

We have explicitly evaluated the corrections implied by two CPT-odd nonminimal couplings to the decay rates of the electroweak processes, $Z_0 \to \bar{l} + l$ and $W^- \to l + \bar{\nu}_l$ $\left(W^+ \to \bar{l} + \nu_l\right)$. Regarding the experimental imprecision in the measurements, upper limits were set on the magnitude of the LV nonminimal coupling at the level of $10^{-6}\ (\text{GeV})^{-1}$ and $10^{-5}\ (\text{GeV})^{-1}$. Other impacts of these NMC in electroweak phenomena are to be investigated, including diferential decay rates of polarized processes, which could, in principle, yield improved upper bounds. More details about this investigation were recently published [16].

5. References

[1] Colladay D and Kostelecky V A 1997 *Phys. Rev.* D **55** 6760
Colladay D and Kostelecky V A 1998 *Phys. Rev.* D **58** 116002
[2] Kostelecky V A and Mewes M 2009 *Phys. Rev.* D **80** 015020
Mewes M 2012 *Phys. Rev.* D **85** 116012
[3] Kostelecky V A and Mewes M 2013 *Phys. Rev.* D **88** 096006
Schreck M 2014 *Phys. Rev.* D **89** 105019
Schreck M 2014 *Phys. Rev.* D **90** 085025
[4] Myers R C and Pospelov M 2003 *Phys. Rev. Lett.* **90** 211601
Reyes C M, Urrutia L F and Vergara J D 2008 *Phys. Rev.* D **78** 125011
Reyes C M, Urrutia L F and Vergara J D 2009 *Phys. Lett.* B **675** 336
Reyes C M 2010 *Phys. Rev.* D **82** 125036
Reyes C M 2009 *Phys. Rev.* D **80** 105008
[5] Cambiaso M, Lehnert R and Potting R 2012 *Phys. Rev.* D **85** 085023
Agostini B, Barone F A, Barone F E, Gaete P and Helayël-Neto J A 2012 *Phys. Lett.* B **708** 212
[6] Ding Y and Kostelecky V A 2016 *Phys. Rev.* D **94** 056008
[7] Casana R, Ferreira Jr. M M, Passos E, dos Santos F E P and Silva E O 2013 *Phys. Rev.* D **87** 047701
Araujo J B, Casana R and Ferreira Jr. M M 2015 *Phys. Rev.* D **92** 025049
Araujo J B, Casana R and Ferreira Jr. M M 2016 *Phys. Lett.* B **760** 302
[8] Altschul B 2011 *Phys. Rev.* D **84** 091902(R)
Altschul B 2013 *Phys. Rev.* D **87** 096004
Altschul B 2013 *Phys. Rev.* D **88** 076015
[9] Noordmans J P, Wilschut H W and Timmermans R G E 2013 *Phys. Rev.* C **87** 055502
Noordmans J P, Wilschut H W and Timmermans R G E 2013 *Phys. Rev. Lett.* **111** 171601
[10] Vos K K, Wilschut H W and Timmermans R G E 2015 *Phys. Rev.* C **92** 052501(R)
Vos K K, Wilschut H W and Timmermans R G E 2015 *Phys. Rev.* C **91** 038501
[11] Vos K K, Noordmans J P, Wilschut H W and Timmermans R G E 2014 *Phy. Lett.* B **729** 112
[12] Castro-Medina J, Novales-Sanchez H and Toscano J J 2015 *Int. J. Mod. Phys.* A **30** 1550216
[13] Aranda J I, Ramirez-Zavaleta F, Rosete D A, Tlachino F J, Toscano J J and Tututi E S 2014 *J. Phys. G: Nucl. Part. Phys.* **41** 055003

IOP Conf. Series: Journal of Physics: Conf. Series **952** (2018) 012019 doi:10.1088/1742-6596/952/1/012019

Aranda J I, Ramirez-Zavaleta F, Rosete D A, Tlachino F J, Toscano J J and Tututi E S 2014 *Int. J. Mod. Phys.* A **29** 1450180
[14] López-Osorio M A, Martínez-Pascual E and Toscano J J 2016 *J. Phys. G: Nucl. Part. Phys.* **43** 025003
[15] Schael S et al. [ALEPH, DELPHI, L3, OPAL, SLD, LEP Electroweak Working Group, SLD Electroweak Group and SLD Heavy Flavour Group Collaborations] 2006 *Phys. Rept.* **427** 257
[16] Mouchrek-Santos V E, Ferreira Jr. M M 2017 *Phys. Rev.* D **95** 071701(R)

IOP Conf. Series: Journal of Physics: Conf. Series **952** (2018) 012020 doi:10.1088/1742-6596/952/1/012020

Tests of Lorentz and CPT symmetry with hadrons and nuclei

J P Noordmans

CENTRA, Departamento de Física, Universidade do Algarve, 8005-139 Faro, Portugal

E-mail: jnoordmans@ualg.pt

Abstract. Using chiral-perturbation-theory techniques, we derive the low-energy effective Lagrangian in terms of pions and nucleons that corresponds to a selected set of dimension-five Lorentz- and CPT-violation quark and gluon operators. The form of the effective operators is determined by the symmetry properties of the original Lagrangian. Using the pion-nucleon Lagrangian, we find the Lorentz-violating contributions to comagnetometer experiments. This results in stringent limits on some of the parameters. For some other parameters we find that the best bounds will come from nucleon-nucleon interactions, and we derive the relevant nucleon-nucleon potential. These considerations imply possible new opportunities for spin-precession experiments involving for example the deuteron.

1. Introduction

It is generally accepted that general relativity and quantum mechanics should become unified around an energy scale around the Planck mass: $m_P \simeq 10^{19}$ GeV. The nature of this unification remains one of the main open issues in theoretical physics today. One of the main difficulties is that m_P, at which deviations from conventional physics should become large, lies far out of reach of any present-day experiments. A possible solution to this problem is to study violations of spacetime symmetries. Many theories that attempt to unify the standard model of particle physics with general relativity contain the possibility for mechanisms that violate symmetries such as Lorentz and CPT symmetry. Moreover, it turns out that very sensitive experimental tests are possible to look for such violations. These considerations have motivated a large amount of research over the last two decades.

This research is facilitated by an effective-field-theory (EFT) approach to Lorentz and CPT violation, called the standard-model extension (SME) [1]. The operators in the Lagrangian of the SME are built from the conventional standard-model quantum fields that represent all the known elementary particles and transform under the Lorentz and gauge group in the conventional way. The only restrictions on the form of the operators that are relaxed, with respect to the standard model, are those of Lorentz and CPT symmetry. Operationally this means that Lorentz indices are no longer contracted to each other, but instead can be contracted to coefficients that parametrize the Lorentz and CPT violation and whose values originate from some unknown high-energy theory that incorporates (spontaneous) Lorentz-symmetry breaking. It is these coefficients that one tries to bound in experiments [2].

One of the SME sectors where limits are relatively weak, is the QCD sector of the SME [2]. This is caused by the nonperturbative nature of QCD at energies below 1 GeV, which prevents

IOP Conf. Series: Journal of Physics: Conf. Series **952** (2018) 012020 doi:10.1088/1742-6596/952/1/012020

precise hadron bounds to be translated to bounds on quark and gluon parameters. Consequently, we started to apply chiral-perturbation-theory techniques to the QCD sector of the SME [3] (See also Refs. [4, 5]). In Ref. [3] we focused on a set of dimension-five quark-gluon operators, which are relevant at an energy scale of about 1 GeV. They were first written down in Ref. [6] and are given by

$$\mathcal{L}^{\mathrm{LV}} = \sum_{q=u,d} \left[C^q_{\mu\nu\rho}\bar{q}\gamma^\mu G^{\rho\nu} q + D^q_{\mu\nu\rho}\bar{q}\gamma^\mu\gamma^5 G^{\rho\nu} q\right] + H_{\mu\nu\rho}\,\mathrm{Tr}\left(G^{\mu\lambda}D^\nu \tilde{G}^\rho{}_\lambda\right) . \tag{1}$$

The sum in this equation runs over the lightest two quark flavors, the fields q are the quark fields and $G^{\mu\nu} = t^a G^{a,\mu\nu}$ ($t^a = \frac{1}{2}\lambda^a$, $a = 1, \ldots, 8$, where λ^a are the Gell-Mann matrices, are the generators of the $SU(3)$ color group). All three operators violate Lorentz symmetry as well as CPT symmetry. The coefficients $C^q_{\mu\nu\rho}$, $D^q_{\mu\nu\rho}$, and $H_{\mu\nu\rho}$ parametrize the Lorentz- and CPT-symmetry breaking. The $C^{\mu\nu\rho}$ and $D^{\mu\nu\rho}$ coefficients are symmetric in their first and last Lorentz indices and antisymmetric in their last two indices, while $H^{\mu\nu\rho}$ is fully symmetric, i.e.

$$X^q_{\mu\nu\rho} = X^q_{\rho\nu\mu} = -X^q_{\mu\rho\nu} , \tag{2}$$

$$H_{\mu\nu\rho} = H_{\mu\rho\nu} = H_{\nu\mu\rho} , \tag{3}$$

with $X \in \{C, D\}$.

There are many more operators that can be considered in the context of chiral perturbation theory, however these are left for future considerations (see for example Refs. [4] and [5]). Also, we consider only the lightest two quark flavors. In principle the strange quark can be included and our work can be extended to $SU(3)$ chiral perturbation theory (e.g. see Ref. [5]).

2. The Lorentz-violating chiral-perturbation-theory Lagrangian

Chiral perturbation theory [7, 8] (see e.g. Refs. [9, 10] for reviews) is used to describe the low-energy ($\lesssim 1\,\mathrm{GeV}$) interactions between pions, nucleons, and photons, which are the relevant degrees of freedom at those energies. These interactions are induced by quark and gluon operators, such as the ones in Eq. (1). The form of the low-energy operators is restricted by the symmetry properties of the original Lagrangian. In Ref. [3] we constructed the low-energy effective Lagrangian that corresponds to the Lorentz-violating operators in Eq. (1) in the $SO(4)$ approach to chiral perturbation theory. In constructing the Lagrangian, one makes use of the fact that the massless (and Lorentz-symmetric) two-flavor QCD Lagrangian is invariant under under chiral transformations of the quark fields, displaying an $SU(2)_L \times SU(2)_R \sim SO(4)$ symmetry. Experimentally, the structure of the baryon and light-meson spectrum implies that the axial part of this symmetry is spontaneously broken at the chiral breaking scale: $\Lambda_\chi \simeq 1$ GeV. The pions are identified as the pseudoscalar Goldstone bosons associated with this spontaneous symmetry breaking, while their small nonzero mass originates from quark masses that form a (small) explicit breaking of the chiral symmetry of the QCD Lagrangian. Such operators that break chiral symmetry can be incorporated in the formalism as well. The way the operators, relevant above 1 GeV, transform under chiral transformations, determines the form of the chiral-breaking operators in the effective Lagrangian below 1 GeV.

Similar to the conventional case, it is possible to construct the low-energy effective operators corresponding to Eq. (1), in terms of the relevant degrees of freedom, i.e. the pions and the nucleons (see Ref. [3] for details). Using the chiral properties of the operators in Eq. (1), together with their C, P, and T characteristics, one finds that the set of lowest-order relativistic operators that is non-redundant under the equations of motion, is given by

$$\begin{aligned}\mathcal{L}_\chi^{\rm LV} &= \frac{i}{m_N}\bar{N}\left(\tilde{C}^+_{\mu\nu\rho}+\tilde{C}^-_{\mu\nu\rho}\left[\tau_3-\frac{2}{F_\pi^2 D}\left(\boldsymbol{\pi}^2\tau_3-\pi_3\boldsymbol{\tau}\cdot\boldsymbol{\pi}\right)\right]\right)\sigma^{\nu\rho}\mathcal{D}^\mu N\\ &+\frac{1}{m_N^2}\tilde{H}_{\mu\nu\rho}\bar{N}\gamma^\mu\gamma^5\mathcal{D}^\nu\mathcal{D}^\rho N+\frac{i}{m_N F_\pi D}\tilde{D}^-_{\mu\nu\rho}\bar{N}(\boldsymbol{\tau}\times\boldsymbol{\pi})_3\sigma^{\nu\rho}\mathcal{D}^\mu N+\\ &+\frac{1}{m_N^2 F_\pi}\breve{D}^+_{\mu\nu\rho\alpha\beta}\bar{N}(\boldsymbol{\tau}\cdot D^\mu\boldsymbol{\pi})\sigma^{\nu\rho}\mathcal{D}^\alpha\mathcal{D}^\beta N+\text{H.c.}\ ,\end{aligned} \tag{4}$$

where $X^\pm_{\mu\nu\rho}=X^u_{\mu\nu\rho}\pm X^d_{\mu\nu\rho}$, with $X\in\{C,D\}$. Explicit expressions for the pion and nucleon covariant derivatives can be found in Ref. [3] and D in the denominator of the third and the fifth term is given by $D=1+\boldsymbol{\pi}^2/F_\pi^2$. The parameter $\breve{D}^+$ is given by

$$\breve{D}^+_{\mu\rho\nu\alpha\beta}=\tilde{D}^{+,1}_{\mu\rho\nu}\eta_{\alpha\beta}+\tilde{D}^{+,2}_{\alpha\nu\rho}\eta_{\mu\beta}+\tilde{D}^{+,3}_{\alpha[\beta\rho]}\eta_{\mu\nu}\ , \tag{5}$$

where square brackets mean antisymmetrization over the enclosed indices. In Eqs. (4) and (5), the Lorentz-violating coefficients at the hadronic level are denoted by a tilde. This represents the fact that they are proportional to the Lorentz-violating coefficients in Eq. (1), but additionally contain an unknown low-energy constant, i.e. $\tilde{C}^\pm_{\mu\nu\rho}=c^\pm C^\pm_{\mu\nu\rho}$, $\tilde{D}^-_{\mu\nu\rho}=d^-D^-_{\mu\nu\rho}$, $\tilde{D}^{+,i}_{\mu\nu\rho}=d_i^+D^+_{\mu\nu\rho}$, and $\tilde{H}^{\mu\nu\rho}=hH^{\mu\nu\rho}$, with $c^\pm$, d^-, d_i^+, and h the low-energy constants. The value of these low-energy constants cannot be determined using symmetry arguments, although chiral symmetry does imply that $d^-=2c^-$. We use naive dimensional analysis (NDA) [11] to get an estimate of the value of the low-energy constants at the order-of-magnitude level. NDA gives that $c^\pm,d^-=\mathcal{O}(\Lambda_\chi F_\pi)$, $d_i^+=\mathcal{O}(F_\pi)$, and $h=\mathcal{O}(\Lambda_\chi^2)$.

An important observation is that in contrast to the $C^\pm_{\mu\nu\rho}$ and $H_{\mu\nu\rho}$ parameters, the $D^\pm_{\mu\nu\rho}$ coefficient does not give any kinetic nucleon term. Since it turns out that the strongest constraints come from properties of the free nucleon field, bounds on $D^\pm_{\mu\nu\rho}$ will be much weaker than those on $C^\pm_{\mu\nu\rho}$ and $H_{\mu\nu\rho}$.

3. Constraints

In Ref. [3] we derived bounds on $C_{\mu\nu\rho}$ and $H_{\mu\nu\rho}$ from comagnetometer experiments. In finding the particular components to which the experiments are sensitive, we followed the approach of Ref. [12]. We will sketch this approach here briefly and refer to Ref. [3] or [12] for more details.

Essentially, one calculates the shift of the rotational energy levels of an atom, ion, or nucleus W, that follows from the Lorentz-violating Hamiltonian described in the previous section. The part of the Hamiltonian for W that is linear in the Lorentz-violating coefficients, is given by

$$h'_W=\sum_w\sum_{N=1}^{N_{w,W}}\delta h_{w,N}\ , \tag{6}$$

where $\delta h_{w,N}$ is the Hamiltonian for the N-th particle of species w, described in the previous section. The second sum runs over all particles of species w that are present in W.

This Hamiltonian induces a shift in the energy levels corresponding to

$$\delta E(F,M_F)=\langle F,M_F|h'_W|F,M_F\rangle\ , \tag{7}$$

where $|F,M_F\rangle$ is the state of W with spin angular momentum F and projection M_F. The Lorentz-violating contribution to a transition frequency $F,M_F\to F',M'_F$ can then be calculated by $\delta\omega=\delta E(F,M_F)-\delta E(F',M'_F)$

Table 1. Order-of-magnitude bounds on the LV tensor components defined in Eqs. (1).

Tensor component	Limit in GeV^{-1}
H_{TTX}, H_{TTY}, $C^q_{T[XZ]}$, $C^q_{T[YZ]}$	$< 10^{-33}$
H_{TTZ}	$< 10^{-30}$
$C^q_{T[XY]}$	$< 10^{-29}$
H_{TXX}, H_{TXY}, H_{TYY}	$< 10^{-28}$
H_{TTT}, H_{TXZ}, H_{TYZ}, $C^q_{T[TZ]}$, $C^q_{X[YZ]}$, $C^q_{Y[XZ]}$, $C^q_{Y[YZ]}$	$< 10^{-27}$
$C^q_{T[TX]}$, $C^q_{T[TY]}$, $C^q_{T[TZ]}$, $C^q_{X[XZ]}$, $C^q_{Z[XZ]}$, $C^q_{Z[YZ]}$	$< 10^{-26}$

In Ref. [3] we found that the a dipole shift is the dominant LV contribution to the energy. It is given by

$$\delta E(F, M_F) = -\frac{2M_F}{F} \sum_{w} \sum_{N=1}^{N_w} \left[H^{300} + (\tilde{C}^w)^{012} - (\tilde{C}^w)^{021} \right] \langle [\sigma_3]_{w,N} \rangle \ , \tag{8}$$

Here, $\langle [\sigma_3]_{w,N} \rangle$ is an unknown matrix element of the third spin component, defined in the 'stretched' state $|F, F\rangle$ (the 3-axis is the quantization axis). One has to adopt some nuclear-structure model to calculate the values of such matrix elements.

The best bounds on the Lorentz-violating dipole contribution in Eq. (8) come from a $^3\text{He}/^{129}\text{Xe}$ comagnetometer [13] for the X and Y direction and from a $^{199}\text{Hg}/^{133}\text{Cs}$ comagnetometer [14] for the Z direction. Here the X, Y, and Z direction are defined in the Sun-centered inertial reference frame [2]. These experiments allow us to put bounds on the LV components in the order of 10^{-33}–10^{-30} GeV^{-1}. Additional bounds can be derived by considering effects due to the nonzero velocity of the Earth. Such effects are ignored in most cases, since they are suppressed by powers of the Earth's velocity with respect to the Sun, i.e. powers of $\beta_\oplus \simeq 10^{-4}$. However, due to the high precision of the comagnetometer experiments, in this case relevant bounds can be derived from such effects. An analysis of such boost-dependent signals was performed in Ref. [15]. The resulting order-of-magnitude limits are collected in table 1.

4. The $D^\pm_{\mu\nu\rho}$ parameter

As can be seen from Eq. (4), the Lorentz-violating coefficient $D^\pm_{\mu\nu\rho}$ does not contribute to any nucleon two-point function, due to its symmetry properties. As such it does not contribute to free nucleon properties and thus cannot be bounded to the same precision as $C^\pm_{\mu\nu\rho}$ and $H_{\mu\nu\rho}$, using comagnetometer experiments.

In Ref. [3], we therefore calculated the pion-loop contribution of $D^\pm_{\mu\nu\rho}$ to the nucleon electromagnetic form factor (see Ref. [3] for the result). The problem with using the electromagnetic nucleon form factor to bound Lorentz-violating coefficients is the fact that the presence of physical electromagnetic fields suppresses the effects by many order of magnitudes. The suppression factor with respect to the dominant Lorentz-violating effects due to C and H will be in the order of $eF^{\mu\nu}/\Lambda_\chi^2$. For a magnetic field of 1 tesla this is about 10^{-16}. We thus estimate that the bounds on $D^-_{\mu\nu\rho}$ from comagnetometer experiments are at best in the order of $\mathcal{O}(10^{-17})$ GeV.

A more promising option is the contribution of $D^\pm_{\mu\nu\rho}$ to nucleon-nucleon interactions. Although it is not possible to write a two-point function for spin-$\frac{1}{2}$ particles involving $D^\pm_{\mu\nu\rho}$, this is not necessarily true for higher-spin particles, such as the deuteron. In Ref. [3] we thus calculated the Lorentz-violating nucleon-nucleon potential that follows from one-pion exchange between nucleons in the context of chiral EFT (see Refs. [16, 17, 18] for reviews). In the relevant pion-exchange diagram, one of the two interaction vertices is a Lorentz-violating vertex. The resulting potential is given by

$$\begin{aligned} V_{\mathrm{LV}} &= -\left(\epsilon^{ijk}\tilde{D}^{-}_{0ij}\right)\frac{2ig_A}{F_\pi^2}(\boldsymbol{\tau}_1\times\boldsymbol{\tau}_2)_3\frac{(\boldsymbol{\sigma}_1\cdot\boldsymbol{k})\sigma_2^k+(\boldsymbol{\sigma}_2\cdot\boldsymbol{k})\sigma_1^k}{\boldsymbol{k}^2+m_\pi^2} \\ &\quad -\left(\epsilon^{jkl}\breve{D}^{+}_{ijk00}\right)\frac{4g_A}{F_\pi}\boldsymbol{\tau}_1\cdot\boldsymbol{\tau}_2\frac{(\sigma_1^l\sigma_2^m+\sigma_1^m\sigma_2^l)k^ik^m}{\boldsymbol{k}^2+m_\pi^2}, \end{aligned} \tag{9}$$

where $\boldsymbol{\sigma}_{1,2}$ ($\boldsymbol{\tau}_{1,2}$) are the spin (isospin) operators of nucleon 1 and 2 and the momentum transfer $\boldsymbol{k}=\boldsymbol{p}-\boldsymbol{p}'$ flows from nucleon 1 to nucleon 2; $\boldsymbol{p}$ and $\boldsymbol{p}'$ are the relative momenta of the incoming and outgoing nucleon pair in the center-of-mass frame. The Latin indices $i, j, k, ...$ denote spatial directions. There are other contributions from the D-parameter to the potential, which can be found in Ref. [3].

The detailed study of the effects of Eq. (9) are left for future work. However, it seems clear that it will have an effect on for example the spin-precession frequency of the deuteron as well as on other light-nucleon systems or clock-comparison experiments. For such systems a bound on $D^{\pm}_{\mu\nu\rho}$ that is a significant improvement over the current bounds seems feasible.

5. Conclusion and outlook

In Ref. [3], we derived the hadronic low-energy effective Lagrangian for Lorentz violation, corresponding to the operators in Eq. (1). Our approach is based on chiral-symmetry considerations and can be extended to all operators of the Lorentz-violating standard-model extension. Steps towards this goal have been taken for several different operators in other works [4, 5].

Using the low-energy effective Lagrangian, we derived the resulting shift in the rotational energy levels of atoms. Using results from comagnetometer experiments [13, 14] we subsequently derived strict bounds on components of two of the three Lorentz-violating coefficients, which are collected in table 1. It is clear that many of the components of $C_{\mu\nu\rho}$ and $H_{\mu\nu\rho}$ have bounds that are well beyond the Planck scale.

In contrast, the bounds on the $D_{\mu\nu\rho}$ coefficients are much less strict, due to that fact that they do not contribute to the nucleon two-point function and therefore cannot be limited by the same methods that provided the stringent bounds on $C_{\mu\nu\rho}$ and $H_{\mu\nu\rho}$. In Ref. [5] we found that this also holds for a particular pure-gluon parameter in the mSME and we expect there will be more such parameters in the SME Lagrangian.

It is likeley that the bounds on $D_{\mu\nu\rho}$ can be improved by considering effects of the one-pion exchange nucleon-nucleon potential in Eq. (9). These effects could provide much better bounds than electromagnetic interactions. Especially the spin precession of the deuteron is promising in this respect and for example storage-ring experiments might be able to place stringent constraints [19].

Acknowledgments

The author acknowledges the financial support of the Portuguese Foundation for Science and Technology (FCT) under grant SFRH/BPD/101403/2014 and program POPH/FSE.

References

[1] Colladay D and Kostelecký V A 1997 *Phys. Rev.* D **55** 6760
Colladay D and Kostelecký V A 1998 *Phys. Rev.* D **58** 116002

[2] Kostelecký V A and Russell N 2016 Data Tables for Lorentz and CPT Violation *Preprint* 0801.0287v10 [hep-ph]
[3] Noordmans J P, de Vries J and Timmermans R G E 2016 *Phys. Rev.* C **94** 025502
[4] Kamand R, Altschul B and Schindler M R 2017 *Phys. Rev.* D **95** 056005
[5] Noordmans J P 2017 *Phys. Rev.* D **95** 075030
[6] Bolokhov P A and Pospelov M 2008 *Phys. Rev.* D **77** 025022
[7] Weinberg S 1979 *Physica* A **96** 327
[8] Gasser J and Leutwyler H 1984 *Ann. Phys.* **158** 142
[9] Weinberg S 1996 *The Quantum Theory of Fields* vol 2 (Cambridge: Cambridge University Press)
[10] Bernard V and Meißner U-G 2007 *Ann. Rev. Nucl. Part. Sci.* **57** 33
Scherer S and Schindler M R 2012 *Lect. Notes Phys.* **830** 1
[11] Manohar A V and Georgi H 1984 *Nucl. Phys.* B **234** 189
Georgi H and Randall L 1986 *Nucl. Phys.* B **276** 241
[12] Kostelecký V A and Lane C D 1999 *Phys. Rev.* D **60** 116010
[13] Allmendinger F, Heil W, Karpuk S, Kilian W, Scharth A, Schmidt U, Schnabel A, Sobolev Yu and Tullney K 2014 *Phys. Rev. Lett.* **112** 110801
[14] Peck S K, Kim D K, Stein D, Orbaker D, Foss A, Hummon M T and Hunter L R 2012 *Phys. Rev.* A **86** 012109
[15] Canè F, Bear D, Phillips D F, Rosen M S, Smallwood C L, Stoner R E, Walsworth R L and Kostelecky V A 2004 *Phys. Rev. Lett.* **93** 230801
[16] Bedaque P F and van Kolck U 2002 *Ann. Rev. Nucl. Part. Sci.* **52** 339
[17] Epelbaum E, Hammer H W and Meißner U-G 2009 *Rev. Mod. Phys.* **81** 1773
[18] de Vries J and Meißner U-G 2016 *Int. J. Mod. Phys.* E **25** 1641008
[19] Eversmann D *et al.* (JEDI Collaboration) 2015 *Phys. Rev. Lett.* **115** 094801

IOP Conf. Series: Journal of Physics: Conf. Series **952** (2018) 012021 doi:10.1088/1742-6596/952/1/012021

CPT and Lorentz violation in the electroweak sector

D Colladay[1], J P Noordmans[2] and R Potting[2]

[1] New College of Florida, 5800 Bay Shore Road, Sarasota, FL 34243, USA
[2] CENTRA and Department of Physics, FCT, University of the Algarve, Campus de Gambelas, 8005-139 Faro, Portugal

E-mail: rpotting@ualg.pt

Abstract. Long ago, Carroll, Field and Jackiw introduced CPT-violation in the photon sector by adding a dimension-3 gauge-invariant term parametrized by a constant four-vector parameter k_{AF} to the usual (Maxwell) Lagrangian, deriving an ultra-tight bound from astrophysical data. Here, we will discuss recent work studying the extension of this term to the full electroweak gauge sector of the Standard Model. In the context of the Standard Model Extension, CPT and Lorentz violation arises from two gauge-invariant terms parametrized by the four vectors k_1 and k_2. First we will show how upon spontaneous breaking of the electroweak gauge symmetry these two terms yield Lorentz-violating terms for the photon and the W and Z bosons. As it turns out, the resulting modified dispersion relations for the W bosons yield spacelike momentum for one of its propagating modes at sufficiently large energy. This in turn allows for the possibility of Cherenkov-like W-boson emission by high-energy fermions such as protons, provoking their decay. Analysis of ultra-high-energy cosmic ray data allows for bounding the previously unbound parameter k_2, and, by combination with the ultra-tight bound on k_{AF}, the parameter k_1.

1. The CPT-odd gauge sector of the minimal Standard Model Extension

In the past two decades there has been a considerable interest in the possibility that Lorentz symmetry may not be exact in nature. The main theoretical motivation for this idea comes from a number of candidate theories for quantum gravity that have been shown to involve Lorentz-invariance violation as a possible effect. This may come about by spontaneous breaking of Lorentz symmetry in theories with Lorentz-invariant dynamics, such as in string field theory [1] or loop quantum gravity [2], or in theories that violate Lorentz invariance at a fundamental level, such as noncommutative geometry [3] (aspects of which are discussed by Orfeu Bertolami in this workshop) or Hořava-Lifshitz gravity [4, 5] (see the talks by Diego Blas and Kevin Grosvenor).

From the experimental point of view, the development of low-energy effective field theories with Lorentz-invariance violation, in particular the Standard-Model Extension (SME) [6], has been instrumental. If this framework, the Lagrangian of the matter sector contains all Lorentz-violating gauge-invariant effective operators that can be build from the conventional Standard-Model fields, coupled to vector and tensor coefficients that parametrize the Lorentz violation. The SME also contains all CPT-violating operators, since in any local interacting quantum field theory CPT violation implies Lorentz violation [7]. The SME can thus be used to provide a general quantification of the exactness of Lorentz and CPT symmetry in the form of observational constraints on the SME coefficients [8].

In 1991, well before the full SME was formulated, Carroll, Field and Jackiw introduced CPT-violation in the photon sector by adding a dimension-3 gauge-invariant term parametrized by a constant four-vector k_{AF} to the usual (Maxwell) Lagrangian [9]:

$$\frac{1}{2}\epsilon_{\mu\nu\rho\sigma}k^{\mu}_{AF}A^{\nu}F^{\rho\sigma}. \tag{1}$$

They managed to extract an ultra-tight bound $|(k_{AF})^{\kappa}| \leq \mathcal{O}\big(10^{-43}\,\text{GeV}\big)$ by analyzing astrophysical data. In the context of the $SU(2)\times U(1)$ electroweak sector of the Standard Model, CPT and Lorentz violation arises from two Chern-Simons-like SME terms parametrized by the four-vectors k_1 and k_2 [6]:

$$\epsilon^{\mu\nu\rho\sigma}k_{1\mu}B_{\nu}B_{\rho\sigma} + \frac{1}{2}\epsilon^{\mu\nu\rho\sigma}k_{2\mu}\big[A^a_{\nu}A^a_{\rho\sigma} + \frac{2}{3}g_2\epsilon^{abc}A^a_{\nu}A^b_{\rho}A^c_{\sigma}\big], \tag{2}$$

where B_μ and A^a_μ (with $a = 1, 2, 3$) represent the $U(1)_Y$ and $SU(2)_L$ gauge fields, $B_{\mu\nu} = \partial_\mu B_\nu - \partial_\nu B_\mu$ and $A^a_{\mu\nu} = \partial_\mu A^a_\nu - \partial_\nu A^a_\mu$, while g_2 denotes the $SU(2)$ coupling constant. Both terms in expression (2) are gauge invariant up to a total derivative. There is a similar term corresponding to the strong-interaction sector, which we will ignore in this paper.

In the Standard Model, spontaneous symmetry breaking

$$SU(2)_L \times U(1)_Y \to U(1)_{EM}, \tag{3}$$

is provoked by assuming a Mexican-hat-type potential for the Higgs field $SU(2)$ doublet $\phi = \begin{pmatrix}\phi^+\\ \phi_0\end{pmatrix}$. In the unitary gauge, ϕ is taken to assume the vacuum expectation value $\frac{1}{\sqrt{2}}\begin{pmatrix}0\\ v\end{pmatrix}$. The quadratic CPT- and Lorentz-violating Lagrangian of the electroweak gauge sector then becomes, in terms of the photon and the W and Z boson fields:

$$\begin{aligned}\mathcal{L}_{AWZ} \;=\;& \frac{1}{2}A_\mu D^{\mu\nu}A_\nu + \frac{1}{2}Z_\mu D^{\mu\nu}Z_\nu + W_\mu D^{\mu\nu}W^*_\nu + \frac{1}{2}m_Z^2 Z_\mu Z^\mu + m_W^2 W_\mu W^{*\mu}\\ &+ \frac{1}{2}\epsilon_{\mu\nu\rho\sigma}\left(k^\mu_{AF}A^\nu F^{\rho\sigma} + k^\mu_{ZZ}Z^\nu Z^{\rho\sigma} + 2k^\mu_{WW}W^\nu W^{*\rho\sigma} + 2k^\mu_{\text{mix}}Z^\nu F^{\rho\sigma}\right),\end{aligned} \tag{4}$$

where $D^{\mu\nu} = \partial^2\eta^{\mu\nu} - (1-\xi^{-1})\partial^\mu\partial^\nu$, $m_W = c_w m_Z = g_2 v/2$ as usual, and

$$k^\mu_{AF} = 2c_w^2 k_1^\mu + s_w^2 k_2^\mu\,, \tag{5}$$
$$k^\mu_{ZZ} = 2s_w^2 k_1^\mu + c_w^2 k_2^\mu\,, \tag{6}$$
$$k^\mu_{WW} = k_2^\mu\,, \tag{7}$$
$$k^\mu_{\text{mix}} = c_w s_w(k_2^\mu - 2k_1^\mu)\,, \tag{8}$$

with $c_w = \cos\theta_w$ and $s_w = \sin\theta_w$. The physical gauge boson fields are defined as usual

$$W_\mu = \frac{1}{\sqrt{2}}\left(A^1_\mu - iA^2_\mu\right), \qquad W^*_\mu = \frac{1}{\sqrt{2}}\left(A^1_\mu + iA^2_\mu\right), \tag{9}$$
$$Z_\mu = c_w A^3_\mu - s_w B_\mu\,, \qquad A_\mu = s_w A^3_\mu + c_w B_\mu\,. \tag{10}$$

2. Dispersion relations in the photon–Z-boson sector

The mixing term between the photon and the Z boson in (4) presents a considerable complication in interpreting the quadratic Lagrangian, as the asymptotic propagating modes no longer correspond exactly to the photon and Z-boson fields, but rather linear combinations thereof (and their derivatives).

As the existing bounds on k_{AF} are so stringent we will assume k_{AF} can be neglected with respect to k_{ZZ}, k_{WW} and k_{mix}. It is valid to do so as long as the bounds we consider on the

IOP Conf. Series: Journal of Physics: Conf. Series **952** (2018) 012021 doi:10.1088/1742-6596/952/1/012021

latter parameters remain much larger than the bounds on k_{AF}. In that case, we see from (5) that $k_1^\mu = -\frac{1}{2}\tan^2(\theta_w)k_2^\mu$ such that $k_{\text{mix}}^\mu = \frac{1}{2}\tan(2\theta_w)k_{ZZ}^\mu$. We now define the eight-component bi-four-vector $\mathcal{A}$, given by

$$\mathcal{A}^\mu = \begin{pmatrix} A^\mu \\ Z^\mu \end{pmatrix} . \tag{11}$$

Lagrangian (4), when restricted to the photon and Z boson fields, then becomes:

$$\mathcal{L}_{AZ} = \frac{1}{2}\mathcal{A}_\mu^T D^{\mu\nu}\mathcal{A}_\nu + \frac{1}{2}\mathcal{A}_\mu^T \mathcal{M}\mathcal{A}^\mu - \mathcal{A}_\mu^T \mathcal{X} k_{ZZ}^{\mu\nu}\mathcal{A}_\nu , \tag{12}$$

where $k_{ZZ}^{\mu\nu} = \epsilon^{\mu\nu\rho\sigma}(k_{ZZ})_\rho\partial_\sigma$, and

$$\mathcal{M} = \begin{pmatrix} 0 & 0 \\ 0 & m_Z^2 \end{pmatrix} , \tag{13a}$$

$$\mathcal{X} = \begin{pmatrix} 0 & \frac{1}{2}\tan(2\theta_w) \\ \frac{1}{2}\tan(2\theta_w) & 1 \end{pmatrix} . \tag{13b}$$

Passing to momentum space, the equations of motion corresponding to Lagrangian $\mathcal{L}_{AZ}$ become

$$\left[(p^2 - \mathcal{M})\eta^\mu{}_\nu - (1-\xi^{-1})p^\mu p_\nu - 2i\mathcal{X}\epsilon^{\alpha\beta\mu}{}_\nu(k_{ZZ})_\alpha p_\beta\right]\tau_\sigma^{(\lambda)\nu} \equiv \mathcal{S}^\mu{}_\nu\tau_\sigma^{(\lambda)\nu} = 0 \tag{14}$$

where $\tau_\sigma^{(\lambda)\nu}(\vec{p})$, with $\sigma = \pm 1$ and $\lambda \in \{0, 3, +, -\}$, represent the eight eigenvectors of the equation-of-motion operator $\mathcal{S}^\mu{}_\nu$.

It can be shown [10] that the corresponding eigenvalues $\Omega_\sigma^\lambda(p)$ of $\mathcal{S}^\mu{}_\nu$ are given by

$$\Omega_{+1}^0(p) = \frac{1}{\xi}p^2 , \tag{15a}$$

$$\Omega_{-1}^0(p) = \frac{1}{\xi}(p^2 - \xi m_Z^2) , \tag{15b}$$

$$\Omega_{+1}^3(p) = p^2 , \tag{15c}$$

$$\Omega_{-1}^3(p) = p^2 - m_Z^2 , \tag{15d}$$

$$\Omega_{+1}^\pm(p) = p^2 - \frac{1}{2}m_Z^2 \pm \delta(p) + \frac{1}{2}\sqrt{(m_Z^2 \mp 2\delta(p))^2 + 4\tan^2(2\theta_w)\delta(p)^2} , \tag{15e}$$

$$\Omega_{-1}^\pm(p) = p^2 - \frac{1}{2}m_Z^2 \pm \delta(p) - \frac{1}{2}\sqrt{(m_Z^2 \mp 2\delta(p))^2 + 4\tan^2(2\theta_w)\delta(p)^2} . \tag{15f}$$

with $\delta(p) = \sqrt{(p\cdot k_{ZZ})^2 - p^2 k_{ZZ}^2} = \frac{2}{\tan(2\theta_w)}\sqrt{(p\cdot k_{\text{mix}})^2 - p^2 k_{\text{mix}}^2}$. The conditions $\Omega_\sigma^\lambda(p) = 0$ define the dispersion relations of the corresponding polarization modes $\tau_\sigma^{(\lambda)\nu}(\vec{p})$.

For "small" energies we can expand the square roots of the final two expressions and obtain

$$\Omega_{+1}^\pm(p) = p^2 + \tan^2(2\theta_w)\frac{\delta(p)^2}{m_Z^2} \pm 2\tan^2(2\theta_w)\frac{\delta(p)^3}{m_Z^4} + \cdots , \tag{16a}$$

$$\Omega_{-1}^\pm(p) = p^2 - m_Z^2 \pm 2\delta(p) - \tan^2(2\theta_w)\frac{\delta(p)^2}{m_Z^2} \mp 2\tan^2(2\theta_w)\frac{\delta(p)^3}{m_Z^4} + \cdots . \tag{16b}$$

From Eqs. (15) and (16) it is clear that the $\sigma = +1$ modes are massless and the $\sigma = -1$ modes are massive, at least in the limit of small Lorentz violation and low energies. We can therefore identify the former mode with the photon and the latter mode with the Z boson.

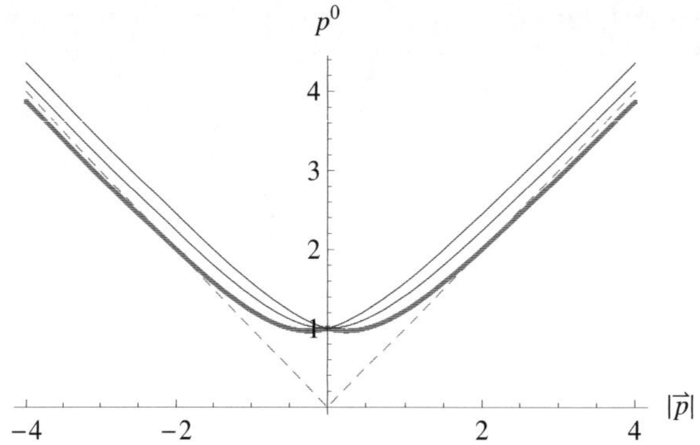

Figure 1. The dispersion relations (19) of the W boson for purely timelike k_2^μ, where we chose $k_2^0 = \frac{m_W}{4}$, in arbitrary units. Indicated in red is the + mode, which has spacelike four-momenta above an energy threshold.

It is not difficult to see from the Lorentz-violating dispersion relations (15e) and (15f) that

$$p^2 \begin{cases} \leq 0 & \text{for} \quad \sigma = +1 \\ > 0 & \text{for} \quad \sigma = -1 \end{cases} . \tag{17}$$

This shows that, if $k_{AF} = 0$ and $k_{ZZ} \neq 0$, the photon mode always has spacelike momenta, while the Z-boson momentum is always timelike. Note from (15c) that the longitudinal photon mode ($\lambda = 3$) is purely massless. It is unphysical and decouples, together with the gauge modes $\lambda = 0$, as usual upon applying, e.g., a BRST quantization procedure.

A more detailed analysis of the dispersion relations can be found in [10].

3. Cherenkov-like emission of W bosons

Let us now restrict Lagrangian (4) to the W boson field:

$$\mathcal{L}_W = W_\mu D^{\mu\nu} W_\nu^* + m_W^2 W_\mu W^{*\,\mu} + \epsilon_{\mu\nu\rho\sigma} k_2^\mu W^\nu W^{*\rho\sigma} . \tag{18}$$

It has been shown in [11] that (18) can be consistently quantized, as long as the components of k_2 are smaller than of the order of m_W, which we will of course assume to be the case.

As we will see, the resulting modified dispersion relations for the W bosons yield spacelike momentum for one of its propagating modes at sufficiently large energy. This makes it possible for Cherenkov-like W-boson emission to take place by fermions coupling to the W boson whose momentum exceeds a certain threshold value. In the case of composite fermions, such as protons, such a process will provoke their decay into lower-energy particles. Analysis of ultra-high-energy cosmic ray data allows for bounding the parameter k_2 and, by combination with the ultra-tight bound on k_{AF}, the parameter k_1.

The Lorentz-violating Lagrangian implies the dispersion relations

$$\Lambda_3(p) = p^2 - m_W^2 = 0, \qquad \Lambda_\pm(p) = p^2 - m_W^2 \pm 2\sqrt{(p\cdot k_2)^2 - p^2 k_2^2} = 0 \ , \tag{19}$$

(Here we suppressed the unphysical gauge mode $\lambda = 0$.) These dispersion relations are represented in Fig. 1 for the case of purely timelike k_2^μ. As it turns out, the $\lambda = 3$ and $\lambda = -$ gauge-boson polarization modes are timelike for any momentum. On the other hand, it follows from (19) that the gauge-boson momentum is spacelike for the $\lambda = +$ mode, if and only if

$$(p\cdot k_2)^2 > \frac{1}{4} m_W^4 \ . \tag{20}$$

Since spacelike momenta are a necessary condition for the desired Cherenkov-like processes, we will only consider W bosons in the + polarization mode.

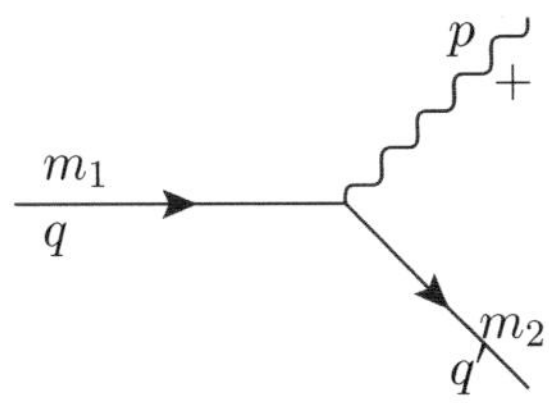

Figure 2. Relevant Feynman diagram for W-boson emission by an incoming fermion.

3.1. Emission by an elementary Dirac fermion

We will first consider the process indicated in Fig. 2 in which an elementary Dirac fermion with a mass m_1 decays to a W boson in the mode $\lambda = +$ and a Dirac fermion with a mass m_2. The Dirac fermions are taken to satisfy regular non-Lorentz-violating dispersion relations. We label the momenta of the particles as follows: the incoming fermion has momentum q, the emitted W boson has momentum p, and the out-going fermion has momentum $q' = q - p$.

Phase-space considerations show that a necessary condition for the decay to take place is

$$4(p \cdot k_2)^2 > (M^2 - (m_1 - m_2)^2)^2 + 4k_2^2(m_1 - m_2)^2. \tag{21}$$

Note that if $m_1 = m_2$ this condition reduces to the space-like condition (20).

The expression for the differential decay rate is

$$d\Gamma = \frac{1}{2q^0}\frac{d^3p}{(2\pi)^3}\frac{1}{\Lambda'_+(p)}\frac{d^3q'}{(2\pi)^3}\frac{1}{2q'^0}\left(\frac{1}{2}\sum_{\text{spins}}|\mathcal{M}|^2\right)(2\pi)^4\delta^4(q-p-q') \,. \tag{22}$$

The squared matrix element $|\mathcal{M}|^2$ is summed (averaged) over the final (initial) fermion spin. The unconventional factor

$$\Lambda'_+(p) = \frac{\partial\Lambda_+(p)}{\partial p^0}, \tag{23}$$

in the denominator defines a positive definite normalization in which the phase space and the matrix element are separately observer Lorentz invariant [11], i.e., invariant under simultaneous Lorentz transformations of the momenta and the Lorentz-violating four-vector. Note that this would not be the case if we were to choose a more conventional phase space normalization.

The matrix element corresponding to the process denoted in Fig. 2 is

$$i\mathcal{M} = \frac{ig_2}{2\sqrt{2}}\bar{u}(q')\gamma^\mu(1-\gamma^5)u(q)e_\mu^{(+)*}(p) \,, \tag{24}$$

where $u(q)$ and $u(q')$ are conventional Dirac spinors. (Analogous expressions can be written down for the case of anti-particles.) The four-vector $e_\mu^{(+)}(p)$ is the gauge-boson polarization vector that corresponds to the dispersion relation in Eq. (19). The explicit expression for the latter can be found in Ref. [11].

The decay rate can be evaluated to yield [12]

$$d\Gamma = -\frac{g_2^2}{64\pi^2 q^0}\int\frac{d^3p}{\Lambda'_+(p)}\theta(q^0-p^0)\delta((q-p)^2-m_2^2)p^2\left[(1\mp X)^2 - \frac{(m_1^2-m_2^2)^2}{p^4}\right] \,. \tag{25}$$

IOP Conf. Series: Journal of Physics: Conf. Series **952** (2018) 012021 doi:10.1088/1742-6596/952/1/012021

where the upper (lower) sign holds for a decaying particle (antiparticle) and

$$X = \frac{p^2(p \cdot k_2 - 2q \cdot k_2) + (m_1^2 - m_2^2)(p \cdot k_2)}{p^2\sqrt{(p \cdot k_2)^2 - p^2 k_2^2}} \,. \tag{26}$$

It follows from formula (25) that for the decay to take place, the incoming fermion momentum has to exceed a threshold value:

$$|\vec{q}| > |\vec{q}|_{\mathrm{th}}, \qquad \text{where} \qquad |\vec{q}|_{\mathrm{th}} \approx \frac{M(M+2m_2)}{2|\kappa|}, \tag{27}$$

where κ is a quantity of the order of the components of k_2: $\kappa \sim \pm|k_2^\mu|$ (see [12] for the exact expressions).

Another result that can be extracted from (25) is that the gauge bosons are emitted in a very narrow forward cone around the direction of the incoming fermion:

$$\cos\theta_{pq} = 1 + \mathcal{O}\left(\frac{\kappa^2}{M^2}\right), \tag{28}$$

where θ_{pq} is the angle between $\vec{p}$ and $\vec{q}$.

Integrating expression (25) over all W-boson momenta yields the total decay rate:

$$\Gamma = \frac{g^2|\kappa|}{64\pi} G(a)\theta(a-1). \tag{29}$$

where

$$\begin{aligned} G(a) = {}& \alpha(a)\left[\frac{1}{a}(-7+3y)\left(1-\frac{m_2}{M}\right) - \frac{1}{a^2}(1-y)\left(1-\frac{3m_2}{M}\right)\right] \\ & - 4\left(1+\frac{1}{a}(1-y)\right)\left(1-\frac{2m_2}{M}\right)\log\left(\frac{1+a-m_2(1-a)/M-\alpha(a)}{1+a-m_2(1-a)/M+\alpha(a)}\right) \\ & + \mathcal{O}\left(\frac{m_{1,2}^2}{M^2}, \frac{\kappa^2}{M^2}\right) \,. \end{aligned} \tag{30}$$

Here a is defined as the ratio of $|\vec{q}|$ to its threshold value, i.e., $a = |\vec{q}|/|\vec{q}|_{\mathrm{th}}$. The function $\alpha(a) = \sqrt{(a-1)^2 + 2m_2(a^2-1)/M}$ and $y = \pm\mathrm{sgn}(\kappa)$, where the upper (lower) sign applies to the particle (antiparticle).

A fermion that couples to a CPT-violating W boson will emit W bosons if it has an energy above threshold, each of which will take away an energy of at least the value $|\vec{q}|_{\mathrm{th}}$. From (29) it follows that the typical decay rate is of the order of 10^{-15} s if $\mathcal{O}(\kappa) = 10^{-7}$ GeV, corresponding to a bound we will find below. It also follows that it will approximately take a time of order $a \times 10^{-15}$ s for all fermions in a decay cascade to fall below threshold for such values of κ.

3.2. Cherenkov emission by a proton

In case the incoming fermion is a composite particle such as the proton, the emission of a W boson will provoke a break-up, since the typical momentum transfer lies in the range of the W-boson mass, which is well within the energy range of for example deep inelastic scattering. In this case, the proton-decay rate can be written as

$$\Gamma = \frac{1}{2q^0}\int \frac{d^3\vec{p}}{(2\pi)^3}\frac{4\pi}{\Lambda'_+(p)} e_\mu^{(+)}(p) e_\nu^{(+)*}(p) W^{\mu\nu} \,, \tag{31}$$

where $W^{\mu\nu}$ is the hadronic part, given by

$$W^{\mu\nu} = \frac{1}{8\pi} \sum_{\sigma} \langle q, \sigma | J^{\nu}(-p) \sum\!\!\!\!\!\!\int_X |X\rangle\langle X| J^{\mu}(p) | q, \sigma \rangle \,. \tag{32}$$

$|q, \sigma\rangle$ is a proton state with momentum q and spin σ, $J^{\mu}(p)$ is the hadronic current, while $\sum\!\!\!\!\!\!\int_X$ indicates a sum over all hadronic final states X along with the corresponding integrations over phase space. $W^{\mu\nu}$ can be evaluated in the parton model (see, e.g., Ref. [13] for a pedagogical introduction).

The final result for the proton decay rate is

$$\Gamma = \frac{g^2|\kappa|}{64\pi} \sum_q \int_0^1 dx \, (f_q(x) + \bar{f}_q(x)) \tilde{G}_q(ax) \theta(ax - 1) \,. \tag{33}$$

Here the functions $f_q(x)$ and $\bar{f}_q(x)$ are the parton distributions functions (PDFs) for the quarks and antiquarks of flavor q, respectively. They represent the chance of finding a quark with momentum fraction x inside the proton. The function $\tilde{G}_q(ax)$ in Eq. (33) is the function in Eq. (30) with the substitutions $m_2 \to x m_2$ and $y \to \tilde{y}_q = \mathrm{sgn}(\kappa) \frac{f_q(x) - \bar{f}_q(x)}{f_q(x) + \bar{f}_q(x)}$. The integral over x in Eq. (33) can be carried out numerically using fits for the PDFs. We refer to Ref. [12] for details.

The most important conclusion for us is that the threshold value for W emission by a proton to occur is still given by formula (27). The decay rate will depend on the details of the PDFs, but its order of magnitude value is the same as for the elementary fermion case (at least for values of a not very close to 1).

4. Limits from ultra-high-energy cosmic rays

We can use the fact that a proton with an energy above threshold will disintegrate to use astrophysical data to limit k_2^{μ}. More precisely, such a proton cannot reach Earth if its mean free path L is much smaller than the distance from its source to Earth. Since many ultra-high-energy cosmic ray particles (UHECR) with energies above $57\,\mathrm{EeV} \equiv |\vec{q}|_{\mathrm{obs}}$ have been observed, more or less from all directions [14], it follows that

$$|\kappa| \lesssim \frac{M^2}{|\vec{q}|_{obs}} \approx 1.1 \times 10^{-7}\,\mathrm{GeV} \equiv |\kappa|_0. \tag{34}$$

We see from (33) that the mean lifetime of protons (in the Earth's frame) t_p is proportional to $|\kappa|^{-1}$. A conservative (large) estimate gives a mean free path of

$$L \simeq c t_p \sim (\hbar c / |\kappa|_0) \times 10^{15} \sim 10^3\,\mathrm{km}\,. \tag{35}$$

It is clear that protons with an energy above this threshold will not be able to reach Earth from any viable UHECR source. This allows us to conclude that

$$|k_2^{\mu}| < 1.1 \times 10^{-7}\,\mathrm{GeV}\,. \tag{36}$$

Combining this with the ultra-tight bound on the components of k_{AF} and its expression (5) in terms of k_1 and k_2, we find the bound

$$|k_1^{\mu}| < 1.7 \times 10^{-8}\,\mathrm{GeV}\,. \tag{37}$$

on the components of k_1.

Acknowledgments

This work is supported in part by the Portuguese Fundação para a Ciência e a Tecnologia (FCT) through projects UID/FIS/00099/2013 and SFRH/BPD/101403/2014 and program POPH/FSE. D. C. acknowledges summer faculty development funds from New College of Florida.

References

[1] Kostelecky V A and Samuel C 1989 *Phys. Rev.* D **39** 683
Kostelecky V A and Potting R 1991 *Nucl. Phys.* B **359** 545
Kostelecky V A and Potting R 1996 *Phys. Lett.* B **381** 89
Hashimoto K and Murata M 2013 *PTEP* **2013** 043B01
[2] Ashtekar A 1986 *Phys. Rev. Lett.* **57**, 2244
Gambini R and Pullin J 1999 *Phys. Rev.* D **59** 124021
[3] Connes A and Kreimer D 1998 *Commun. Math. Phys.* **199** 203
[4] Horava P 2009 *Phys. Rev.* D **79** 084008
[5] Blas D, Pujolas O and Sibiryakov S 2010 *Phys. Rev. Lett.* **104** 181302
[6] Colladay D and Kostelecky V A 1997 *Phys. Rev.* D **55** 6760
Colladay D and Kostelecky V A 1998 *Phys. Rev.* D **58** 116002
[7] Greenberg O W 2002 *Phys. Rev. Lett.* **89** 231602
[8] Kostelecky V A and Russell N 2011 *Rev. Mod. Phys.* **83** 11 (*Preprint* arXiv:0801.0287v10 [hep-ph])
[9] Carroll S M, Field G B and Jackiw R 1990 *Phys. Rev.* D **41** 1231
[10] Colladay D, Noordmans J P and Potting R 2017 *Symmetry* **9** 248
[11] Colladay D, McDonald P, Noordmans J P and Potting R 2017 *Phys. Rev.* D **95** 025025
[12] Colladay D, Noordmans J P and Potting R 2017 *Phys. Rev.* D **96** 035034
[13] Peskin M E and Schroeder D V 1995 *An Introduction to quantum field theory* (Reading, USA: Addison-Wesley)
[14] Aab A et al. [Pierre Auger Collaboration] 2015 *Astrophys. J.* **804** 15
Abu-Zayyad T et al. [Telescope Array Collaboration] 2013 *Astrophys. J.* **777** 88